EAST MIDLANDS GEOTECHNICAL GROUP
THE INSTITUTION OF CIVIL ENGINEERS

Lime Stabilisation

Proceedings of the seminar
held at Loughborough University
Civil & Building Engineering Department
on 25 September 1996

 Thomas Telford

Organisers: The East Midlands Geotechnical Group of the Institution of Civil Engineers

Organising Committee: Dr C. D. F. Rogers, Dept of Civil and Building Engineering, Loughborough University; Dr S. Glendinning, Dept of Civil and Building Engineering, Loughborough University; R. D. Price, Engineering Services Laboratory, Northamptonshire County Council; Dr N. Dixon, Dept of Civil and Structural Engineering, Nottingham Trent University; Dr E. J. Murray, Murray-Rix Geotechnical

Published for the organisers by Thomas Telford Publishing, Thomas Telford Services Ltd, 1 Heron Quay, London E14 4JD

First published 1996
Reprinted 2001

Distributors for Thomas Telford books are
USA: American Society of Civil Engineers, Publications Sales Department, 345 East 47th Street, New York, NY 10017-2398
Japan: Maruzen Co. Ltd, Book Department, 3Ä10 Nihonbashi 2-chome, Chuo-ku, Tokyo 103
Australia: DA Books and Journals, 648 Whitehorse Road, Mitcham 3132, Victoria

A catalogue record for this book is available from the British Library

Classification
Availability: unrestricted
Content: collected papers
Status: authors' opinions
User: civil and geotechnical engineers and landowners

ISBN: 0 7277 2563 7

Printed in Great Britain by The Cromwell Press, Trowbridge, Wiltshire

Acknowledgements

The editors gratefully acknowledge the following:

- Permission to publish Figures 1 and 2 in the paper by H. M. Greaves, which are reproduced from 'Soil stabilisation with cement and lime' by P. T. Sherwood (ISBN 0 11 551171 7), published by HMSO, London. Crown copyright is reproduced with the permission of the Controller of HMSO.

- Permission to publish Figure 5 in the paper by S. Biczysko, which is reproduced from 'Penetration Testing 1988' Proceedings of the First International Symposium ISOPT-1, Orlando, 20–24 March 1988, 1096 pp. two volumes, HFl. 440/£180.00. A. A. Balkema, PO Box 1675, Rotterdam, Netherlands.

- Permission to publish Figures 1 and 2 in the paper by C. D. F. Rogers and S. Glendinning, which is reproduced from 'Site visit report on the construction of a lime modified subgrade on highways' Nos 339-01 and 334-02 by D. A. Sweeney, published by the Geotechnical Engineering Group, Department of Civil Engineering, College of Engineering, University of Saskatchewan, Sasketoon, Saskatchewan, November 1987.

All other material that is referred to is fully referenced in the papers in which the reference is made. Unpublished photographs have been included in the paper by J. H. Smith. These were supplied by the following organisations: Caterpillar Ltd, Wirtgen Ltd and Bomag Ltd. Their provision is also gratefully acknowledged.

Preface

Stabilisation of clay soils using lime is a well tried and tested technique in many overseas countries, but has not been used to its full potential in the UK. The primary application has traditionally been the improvement of clay subgrades for road pavement construction using mix-in-place techniques, although novel applications have emerged to produce a range of geotechnical applications. Associated with the increase in alternative solutions to geotechnical problems is the increase in environmental and economic pressure to reduce the demand for quarried aggregates and disposal of structurally unsuitable or waste materials to landfill sites. The two are not unconnected, and thus any technique that utilises the soil *in situ* must be attractive. Stabilisation using lime is one such technique, regardless of whether the lime is mixed-in-place or introduced by some other means.

A further growing influence in the construction industry is the need for quality assurance and certainty, as far as is possible, that design lives will be met. This mitigates against adoption of new techniques, particularly when the material whose (altered) properties are ultimately to be relied upon to perform well is initially likely to be variable to some degree. However engineers, although being necessarily conservative, also necessarily rely upon the experience of others via technical papers, reports or books. Thus experience of research programmes and/or practical application by others allow newer techniques to be adopted. The pressure for consideration of new techniques derives from the financial (and by implication environmental) demands that an engineer should adopt the most economical solution that can be assured of producing the required performance.

The seminar on Lime Stabilisation and this associated publication aim to meet the need for increased knowledge on a suite of alternative ground improvement techniques using lime. Publication of papers on the subject is particularly important in light of UK experience of the techniques which, although generally good, suffered setbacks with isolated failures including the highly publicised M40 failure. The reasons for these failures are now known and the lessons learnt have been incorporated in current practice, yet publication on the subject has been sparse. Allied with this has been a considerable amount of research effort in the field. The Lime Stabilisation '88 and '90 Seminars organised by BACMI provided an excellent forum for dissemination of such information, but since then there has been no single occasion at which this could be done.

It is for these reasons that the East Midlands Geotechnical Group decided to organise the seminar on the subject in 1996, following an equally successful seminar on Groundwater Pollution in 1984. The EMGG formed four years ago with the aim of providing learned society activities on geotechnical subjects for the benefit of engineers living in the area to either side of the M1 corridor from Northamptonshire to South Yorkshire. Most of the activity concerns the evening meetings programme, although the organisation of site visits and seminars complement this activity. Widening of seminar participation to attract national participation thus provided a logical extension to the aims of the group. The decision to host the seminar at Loughborough University was considered appropriate since it has recently completed ten years of research in the subject area.

The editors wish to acknowledge the considerable support of both the organising committee and the full EMGG committee. They also wish to thank the contributors for their excellent papers, particularly since the time to write such a paper appears to be progressively hard to find. Meeting the deadline for camera ready copy so that the book could be made available at the seminar requires a considerable concentration of effort and this is much appreciated. Additional support has been provided by the members of the Geotechnical Engineering Research Group, notably by Mrs E M Willson and Mrs R Zidan in terms of final production. It is hoped that the next seminar on Lime Stabilisation in the UK will not take another six years to arrange.

Contents

THE LIME STABILISATION PROCESS

Introduction

An Introduction to Lime Stabilisation

H M Greaves BSc (Hons), CEng, MICE.
Stabilisation Services Manager, Buxton Lime Industries.

Construction of Lime and Lime Plus Cement Stabilised Cohesive Soils

J H Smith BSc (Hons), CEng, MICE.
Consultant to Powerbetter Developments Limited.

The Uses of Lime in Ground Engineering: a review of work undertaken at the Transport Research Laboratory

J Perry BSc (Hons), MSc, PhD, CEng, MIMM, FGS.
Civil Engineering Centre, Transport Research Laboratory, Crowthorne.

D J MacNeil BSc.
Civil Engineering Centre, Transport Research Laboratory, Crowthorne.

P E Wilson BSc (Hons), CEng, MICE.
Road Engineering and Environmental Division, Highways Agency, London.

Introduction

Clay soils notoriously provide a challenge to the geotechnical engineer due to their considerable variety in terms of composition and properties, and in particular their variation in properties with time and loading. The latter change in properties is often a manifestation of clay's ability to develop high pore water pressures, both positive and negative (suctions). These pressures influence greatly the stresses that a clay can withstand in engineering practice, whether applied statically or dynamically as a repeated or cyclic loading.

Two broad categories of clay strata are encountered in the UK. Much of the clay exposed at the surface is heavily overconsolidated as a result of erosion subsequent to deposition or glaciation. In their undisturbed state they are generally strong, but when disturbed they usually undergo stress relief, developing negative pore water pressures and hence tending to swell. Good examples of this behaviour are the cases of cutting construction in these clays, embankment construction carried out using these clays and remoulding of the clays for clay (mineral) liners for landfill sites. In addition such clay soils can become impassable or unworkable during construction operations that take place either during or following wet weather. The second broad category concerns soft, wet clayey and silty alluvial deposits, which are typically normally consolidated or lightly overconsolidated. These materials generally require improvement before construction can take place above them.

The properties of a clay soil are largely controlled by the amount and type of clay in the soil, as determined by the Clay Fraction and clay mineralogy respectively, and its stress history. The type of clay mineral present in a clay can affect considerably the way in which it reacts to water. A clay soil containing significant quantities of smectitic minerals (e.g. montmorillonite), for example, will tend to be highly plastic. This is because the mineral has a large capacity for water molecules to become associated with it. The plasticity is measured by the Liquid and Plastic (or Atterberg) Limits, and more precisely the plasticity index being the difference between the two. Thus smectitic clays have a large plasticity index. They also have a high Cation Exchange Capacity (CEC), making them sensitive to the predominant cation present in the soil water. Clay soils containing predominantly kaolinite will have a much lower plasticity (and CEC), and those containing illite will tend to have a lower value still. Thus the simple Atterberg Limit tests provide a good indication of the likely behaviour of a clay soil, and importantly its likely reaction to water, by giving a crude indication of the minerals present.

The influence of stress history has been alluded to above and concerns the clay's consolidated state. The fact that (clay) soils have a memory is a result of their plasticity (or lack of elasticity). This means that consolidation of a clay under a higher level of stress than previously experienced will render it permanently more consolidated, or denser, than prior to the stress application. The stress history is a crucial factor in determining how a clay soil will react to wetting and to imposed loading or unloading, for example by excavation or remoulding *in situ*. The consolidated state of a clay can be *crudely* gauged by observation and simple testing on site, but can only be accurately determined in such equipment as the oedometer or triaxial cell in the laboratory or via sophisticated testing *in situ*.

Improvement of the properties of soils for construction purposes can be achieved by a variety of means, which are broadly covered by

- mechanical improvement, such as densification and drainage,
- chemical improvement, or
- reinforcement by physical inclusion of tensile elements or grouts.

Soil improvement is usually an alternative to the provision of a structural solution to a practical problem, such as provision of retaining walls or buttresses, thickening of road pavements or increasing the extent of foundations. Thus the economics of the alternative solutions will be considered in an engineering design. In the case of clay soils, chemical improvement is commonly most effective since it can be used to change the nature of the material. Chemical means can be used to strengthen the soil, but also to remove its sensitivity both to water and its subsequent stress history. Lime provides an economic and powerful means of chemical improvement, as demonstrated by the dramatic transformation that is evident on mixing lime with a heavy clay. It has not, however, been used to its full potential in the UK. This chapter, and indeed the full publication, aims to provide an introduction to the methods and applications of the lime stabilisation in order to engender greater confidence in the technique.

The first paper, by Greaves, provides a brief history of the subject and a summary of the improvements that can be achieved using lime. Two sets of reactions take place, known as modification and stabilisation reactions, by which the nature of the clay is changed and the modified material becomes cemented respectively. They cause the strength and stiffness of the material to increase markedly and the plasticity, or ductility, to become progressively lost. An important point to note here is that the clay, once modified, can no longer be considered to be a clay and thus it must be treated accordingly as a material in its own right. This material has properties that change with time after the introduction of lime (however achieved) has taken place, the strengthening and stiffening for example being progressive. The permeability of the material, which immediately after treatment could be considered to be a malleable aggregate that progressively cements with time, increases considerably in the short term but progressively decreases as the cementing takes place. The susceptibility to water is greatly reduced, as evidenced by its altered shrinkage and swelling properties and plasticity index. The material becomes more workable since it is apparently drier (it has a typically much increased Plastic Limit) and can be effectively compacted only at a greatly increased water content. The changes in nature are thus considerable.

Greaves goes on to indicate which types of soil are suitable for reaction with lime and the problems associated with excessive sulphates in the soil. Sulphates can affect the reactions by modifying the reaction products, thus demonstrating a need to understand the lime modification and stabilisation processes and to carry out a carefully considered site investigation. Some of the applications of lime improvement are thereafter discussed, together with the opportunities of lime as an environmentally friendly solution to the construction industry's demand for aggregate and apparent need for landfilling.

In the second paper, Smith presents a more detailed review of the type of soil suitable for lime improvement using the mix-in-place technique. He thereafter describes in detail the construction procedures adopted and the safety precautions necessary to ensure that the effects of a strong alkali are not experienced beyond the soil being treated. Research has shown that *appropriate* testing at the site investigation stage is important for accurate design of the process. It is important on site that sufficient water is available to ensure that the modification reactions take place fully and that an adequate period (known as mellowing) is allowed prior to compaction. This is to ensure that the initial reactions, which are expansive in nature, do not affect the compacted state of the mixed material, which should be densely compacted with particles in intimate contact in order for the subsequent stabilisation reactions to be most effective. In this respect addition of water to the material is usually necessary to ensure that a water content at, or slightly wet of, the optimum water content is achieved. The important points to note here are that it is *not* the same as adding water to a clay, since the material is no longer a clay, and that the optimum water content of the modified material will be greatly increased above that of the original clay. These considerations, as Smith points out, are crucial to the success of the mix-in-place process.

The third paper is jointly authored by the Transport Research Laboratory (TRL) and the Highways Agency, and discusses wider applications of lime improvement. The traditional use of lime stabilisation is in the treatment of clay subgrades to create improved road foundations without the need for large quantities of imported granular aggregates. In producing a revised specification for this, the TRL conducted a careful review of the process and the factors affecting it. They provide detailed recommendations for practice, specifically addressing the identification of possible sulphate attack, compaction wet of optimum and the influence of organic matter. They also describe work carried out, or anticipated, in the fields of bulkfill modification of clays, deep stabilisation and processing of contaminated materials for use as fill. All of these applications are the subject of current research projects being carried out at Loughborough University and are described in individual papers later in this book. The paper provides a good introduction to the subjects, and in the case of bulkfill modification valuable complementary data. The application to railway foundations is similar to that of road foundations, yet with important differences since the process must be completed and the track returned to use within hours. It therefore provides another good example of an application in which the two stage process of modification and stabilisation combine to provide both short-term and long-term benefits.

An Introduction to Lime Stabilisation

H.M.GREAVES
Buxton Lime Industries Limited

HISTORY

The use of Lime Stabilisation of clay in construction is over 5,000 years old. The Pyramids of Shersi in Tibet were built using compacted mixtures of clay and lime. China and India have used Lime Stabilisation in various ways throughout their long history.

It was, however, in the USA during the late 1940's that the developing techniques of soil mechanics laboratory testing were applied to evaluating soil-lime mixtures. The treatment of clays with lime started in the 1950's and the technique increased rapidly in popularity. Thousands of miles of state highway and major airports, such as Dallas Fort Worth Airport, were built on lime stabilised clays.

Although laboratory work and limited amounts of site work were carried out in the UK in the 1950's and 1960's, lime stabilisation techniques were not widely used in the UK until the late 1970's. 1956-1962 marked a significant period when investigations into lime and cement stabilisation were undertaken by the Road Research Laboratory, Cement and Concrete Association, Military Engineering Experimental Establishment and the Air Ministry Works Directorate.

Work continued to increase during the early 1980's, mainly in the South East of England and at airports, culminating with a method for the lime stabilisation of subgrades being included in the Department of Transport's Specification for Highway Works" published in 1986. Both lime and cement stabilisation remain options for capping materials in the current revision of the Specification (DTp, 1991).

Lime stabilisation has now been used extensively in the UK with approximately half a million cubic metres of soil being treated in 1995. Despite problems with the M40 in the Spring of 1990, the overwhelming performance of lime treated soils has been good. For even longer term proof of performance, one needs to look at the experience gained in other countries. Kelley (1977) reviewed the use of the process at numerous sites in the USA between 1940 and 1960. He found not only did the process result in savings in construction costs, but the roads and airfields performed well over a 25-30 year period with minimal maintenance.

Considerable work on the site investigation and testing of soils to be stabilised with lime has recently been carried out by the Transport Research Laboratory and the Highways Agency (Perry *et al*, 1995). This has now been incorporated into the Department of Transport Design Manual for Roads and Bridges, HA74/95 and constitutes the most authoritative document on the "Design and Construction of Lime Stabilised Capping".

EFFECT OF LIME ON SOILS

Lime products

Lime stabilisation is achieved with calcium oxide (quicklime or burnt lime) or calcium hydroxide (slaked or hydrated lime). Agricultural lime is usually calcium carbonate and is ineffective for soil improvement and stabilisation. The stabilising effect depends on the reaction between lime and the clay minerals. The main effects of this reaction are:

An increase in the shear strength and bearing capacity of the soil.

A reduction in the susceptibility to swelling and shrinkage.

An improvement in the resistant to trafficking and to bad weather.

A reduction in the moisture content and an improvement in workability and compaction characteristics.

Either quicklime or hydrated lime can be used for soil stabilisation. In the UK quicklime has several advantages over hydrated lime:

Quicklime has a higher available lime content per unit mass than hydrated lime. 3% quicklime is normally equivalent to 4% hydrated lime.

Quicklime is denser than hydrated lime requiring less storage and transport space.

Quicklime is considerably less dusty than hydrated lime.

Quicklime produces a large reduction in moisture content due to hydration and evaporation. It is particularly beneficial with wet soils.

Quicklime generates heat which accelerates strength gain. This is of benefit in a temperate climate such as that found in the UK.

Mechanisms of lime stabilisation

The addition of lime to fine grain soils has several reactions:

Drying out by absorption and evaporation. The reduction in the moisture content of the soil can be substantial and occurs immediately the lime and soil are mixed.

Rapid physio-chemical reactions between the lime and clay minerals produce immediate changes in soil plasticity and workability. This is known as soil improvement or modification.

Long term soil-lime pozzolanic reactions result in the formation of cementing agents which increase strength and durability. This is known as lime stabilisation.

Drying

If quicklime is used for stabilisation rather than hydrated lime, the moisture content of the natural soil can be significantly reduced. Quicklime will immediately take up 32% of its own mass of water from the surrounding soil and slake to form an hydrated lime. The actual moisture content of the soil is, therefore, reduced in addition to the apparent drying out caused by the increase in the Plastic Limit. The slaking reaction of quicklime is highly exothermic and the heat generated causes further water loss due to evaporation. The water loss by this means can be equal to the water loss caused by the slaking of the quicklime.

Improvement/modification

When lime is mixed with soils containing clay minerals in the presence of water, a cation exchange takes place which produces a new material. The soil is transformed to a needle like interlocking metalline structure compared to the plate like structure of natural clays. Hence there are significant changes in the soil's engineering properties.

The effect of the addition of line on the plasticity of London Clay has been presented by Sherwood (1967). The significant increase in Plastic Limit and resulting drop in plasticity index is shown in Figure 1. Sherwood (1967) indicates that the Plastic Limit increases from 24% to 43% with a 4% addition of lime. Beyond this threshold further additions of lime failed to reduce plasticity.

The value of this effect can be appreciated by consideration of the soil shown in Figure 1 at a moisture content of 35%. Since the Plastic Limit of this soil is 25% it is clear that at a moisture content of 35% it will be in a wet and sticky condition, impossible to compact and impossible to traffic. The addition of 2% of lime will change the Plastic Limit to 40% so that the moisture content of the soil will be 5% below the Plastic Limit. Clearly this would greatly enhance the workability and trafficability of this soil.

Perry *et al* (1995) found that lime stabilised soils behave in a very different manner to most naturally occurring soils and have completely different plasticity characteristics to the original unstabilised material. Their research showed that the MCV test is able to reflect these changes. After the mellowing period an MCV-moisture content relationship can be established, although care has to be taken to ensure only the "wet" leg of the graph is used to ensure that adequate compaction is achieved.

Long term reaction - stabilisation

Clay minerals are natural pozzolanas and have the ability to react with lime added to the soil to produce cementitious products. The lime added to the soil results in an increase in the pH to a value in excess of 12 with a resultant increase in the solubility of siliceous and aluminous compounds which react with calcium to form calcium silica hydrates and calcium allumina hydrates. The calcium silicate/aluminate occurs initially in gel form to coat the soil particles, to form a bond, which eventually crystallises into calcium silicate/aluminate hydrate. The cemenititious products are broadly similar in composition to those of cement paste. The process is relatively slow because the available lime has to defuse through the soil structure and the initial cementitious products to the reaction sites. This reaction results in a gain in strength as illustrated in Figure 2.

Most, but not all, clays found in temperate regions are sufficiently reactive for significant strength to develop when they are stabilised with lime. However, at ambient temperatures the reaction is slow and the stabilised soil can continue to gain in strength over a number of years.

SUITABLE SOIL TYPES

As explained previously, clay is a natural pozzolan containing silica and alumina. Free lime reacts with these elements causing gels to form very similar to those found in a cement paste. Other substances such a pulverised fuel ash, blast furnace slag and colliery shale also contain significant amounts of silica and alumina and therefore have the potential to react with lime.

The French use large quantities of PFA-lime mixtures which are very suitable for use as sub-base and road base layers in pavement construction. Some projects are currently being proposed in the UK using these materials and contracts have been undertaken where ground granulated blast furnace slag and lime have been used to stabilise granular materials.

The reaction between lime and clay is dependent on the reactive clay content. Generally the plasticity index is used as a measure of the clay content and Perry *et al* (1995) indicate that a lower limit of 10% should ensure the suitability of the soil for the reaction to take place.

Silts may also be improved with lime to reduce their susceptibility to water, but may not develop much long term pozzolanic strength gain. In such circumstances consideration should be given to using a blend of materials. PFA can be utilised where an additional pozzolanic agent is required to reinforce the silt reaction and obtain the required strength.

THE EFFECT OF SULPHATES

It has long been known that the presence of sulphates in a soil treated with lime may cause problems by reacting with the cementitious materials to cause heaving. This is a complex reaction relying on a number of factors but the products formed in this reaction occupy a greater volume than the constituent parts resulting in loss of strength and volume expansion. Littleton (1995) concluded that contributory factors are:

The type and solubility of the sulphate.

The amount of sulphate present.

The amount and size of clay particles present.

The ability of the soil to absorb water above the level at which the mix was compacted.

This subject is addressed in detail by Perry *et al* (1995) and recommendations have subsequently been published in HA74/95, DTp Design Manual for Roads and Bridges. Providing these recommendations are adhered to and the results of laboratory heave tests are within limits, the lime stabilised material will perform satisfactorily.

It is important to realise that additional sulphates can be brought into the treated material via ground water once construction has been completed. Monitoring of sulphates in ground water is therefore required at the site investigation stage, as it is when any cementitious system is to be buried. Acceptable limits for sulphates in this situation are also specified in the above document.

The ability of ground water to penetrate lime stabilised soil is highly dependent on the permeability and hence the compaction is of fundamental importance. It is essential that the treated soil is compacted with no more than 5% air voids and that close attention is focused on the moisture content at the time of compaction to ensure this is achievable.

LIME STABILISATION IN THE MARKET PLACE

Lime stabilisation has been used regularly throughout the UK since the early 1980s. Although there have been some failures, notably the M40, the extent of these failures has been small. Improved knowledge and testing procedures formulated by a variety of researchers now means that production and control of lime treated soils can be carried out with great assurance.

An analysis of the current lime stabilisation market shows that a small number of contractors and consultants are involved in the majority of contracts. As this system is used purely on the physical or financial benefits it can provide, one can only assume that these companies have learnt how to use lime stabilisation effectively and are reaping the benefits. This means however that a large
number of other such companies are failing to identify the advantages and are possibly putting themselves at a disadvantage.

The natural conservatism of UK engineers now needs to be overcome to fully release the potential that lime offers to the construction industry. Over the last few years, lime stabilisation has often been used as a last resort where other methods have failed. Ironically this has resulted in the softest, wettest and most marginal soils being successfully stabilised often with considerable financial benefits. Not only does lime stabilisation provide opportunities for extremely difficult site conditions to be surmounted but it allows in-situ soils to be used in a whole variety of applications. Consideration of these opportunities needs to be examined during the design and tendering process so that a choice can be made on economic and technical merit rather than as a last resort.

Lime stabilisation of soils should be used in the following situations:

If materials are unacceptably wet or plastic.

Where improved workability and compaction properties are needed.

When greater soil strength and stability is required.

Where off site disposal needs to be avoided.

When materials up to sub-base quality are required from in-situ soils.

Where temporary works and haul roads are needed.

When there is a need to encapsulate difficult materials.

On sites easily affected by adverse weather conditions.

By using stabilisation in the above instances a number of benefits are obvious. It is possible to work with the naturally occurring soils so that excavation and replacement is eliminated. This saves on the import of new materials, transportation costs and more importantly tipping charges.

In some instances treated materials can be used twice. In the case of a piling platform constructed from lime stabilised soil it may be possible to use this layer for the floor slab after piling is complete. If some damage has occurred during piling then remedial work is easily effected as lime treated soil is self healing. Re-rotivation and compaction will return the platform to a suitable state for reuse. Alternatively, if it is not required any further then it can be ploughed up and the trouble of removing the original platform is eliminated.

It should always be remembered that lime treated soils are totally different to the natural material. Strength is enhanced because of the increase in cohesion but the stability of the soil is also enhanced. This makes it possible to improve soils prone to movement by settlement or swelling so that they can be utilised in structural applications.

OPPORTUNITIES FOR THE FUTURE

Opportunities for the use of lime stabilised systems within the construction industry have never been greater. Pressures are continually growing to minimise the use of scarce natural resources and penalties are being introduced on the movement and tipping of excess materials. Lime treated in-situ soils can substantially reduce the impact of such problems.

Increased competition is causing contractors to search harder for economical solutions and systems that can be operated through a variety of weather conditions. The reduced availability of green field sites in the future will also mean that brown field sites will need to be redeveloped more often. In all these aspects improvement with lime can yield significant benefits.

Work is now taking place across Europe to produce European Standards for lime treated materials. This will ensure that the considerable expertise that exists within other European counties will become available in the UK and will provide a common framework for the use of lime treated soils. Greater opportunities will therefore be available to specify these systems and the best of the European experience will be at our disposal.

European countries are much more innovative in their use of secondary materials and initiatives by the TRL and CIRIA will cause the UK to move in the same direction. Many of the systems employed rely on the principles used in lime stabilisation and lime often provides the activator for the binders. This field opens up many alternative construction methods to the designer and the contractor and their use may determine the difference between winning and losing contracts. Government agencies are becoming much more aware of the need to allow these systems to be specified and changes in this area can be expected in the near future. How much more sensible to use cheap waste materials local to each construction site rather than depleting expensive and dwindling natural reserves of stone which need to be hauled over much greater distances.

With the increasing power of the environmental lobby and the opposition by local communities to the disruption caused by construction work, it is time to look closely at alternatives. The use of in-situ materials and locally available secondary materials makes good sense financially and environmentally. Lime plays a key part in both these systems and a thorough understanding of the techniques and opportunities will be essential in the future. Indeed the benefits that derive from the use of lime for the treatment of soils will be so significant that those who do not embrace it will find themselves at a serious disadvantage.

REFERENCES

Department of Transport (1986), "Specification for Highway Works", Sixth Edition, HMSO, London.

Department of Transport (1991), "Specification for Highway Works", Seventh Edition, HMSO, London.

Heath, D C (1992), "The Application of Lime and Cement Soil Stabilisation at BAA Airports" Proc.lnst Civ Engrs Transp, 95, Feb, p.11-49.

Kelley, M (1977), "A Long Range Durability Study of Lime Stabilised Bases at Military Posts in the Southwest", Bulletin 328, National Lime Association WSA.

Littleton, I (1995) "Some Observations on the Presence of Sulphates in Lime Stabilised Clay Soils", Buxton Lime Industries Ltd.

Perry, J, Snowdon, R A, and Wilson, P E (1995), "Site Investigation for Lime Stabilisation of Highway Works", Department of Transport.

Sherwood, P T (1967), "Views of the Road Research Laboratory on Soil Stabilisation in the United Kingdom", Cement Lime and Gravel, Vol 42, No 9, p.277-280.

Snedker, E A, and Temporal, J (1990), "M40 Motorway Banbury IV Contract - Lime Stabilisation", Highways and Transportation, December, p.7-8.

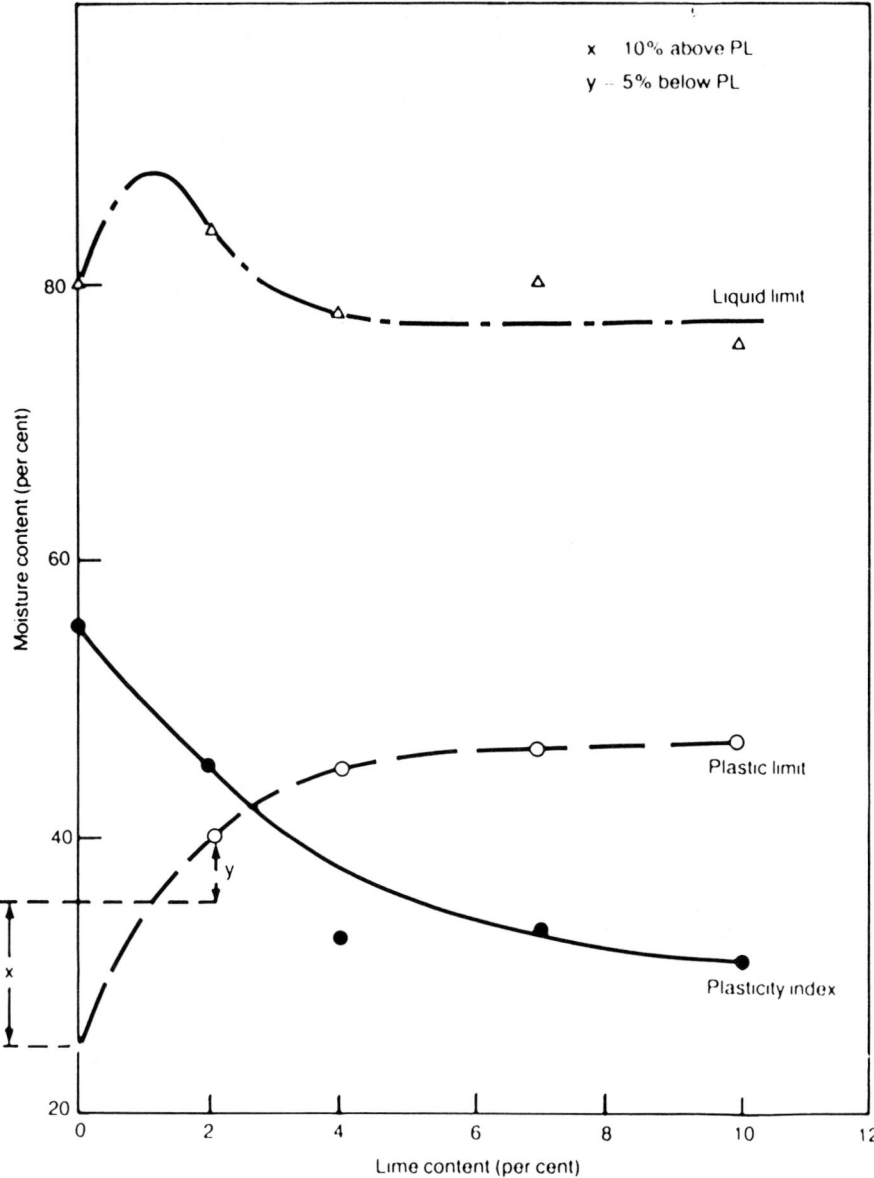

Figure 1   Effect of lime addition on the Plasticity of London Clay (after Sherwood, 1967).

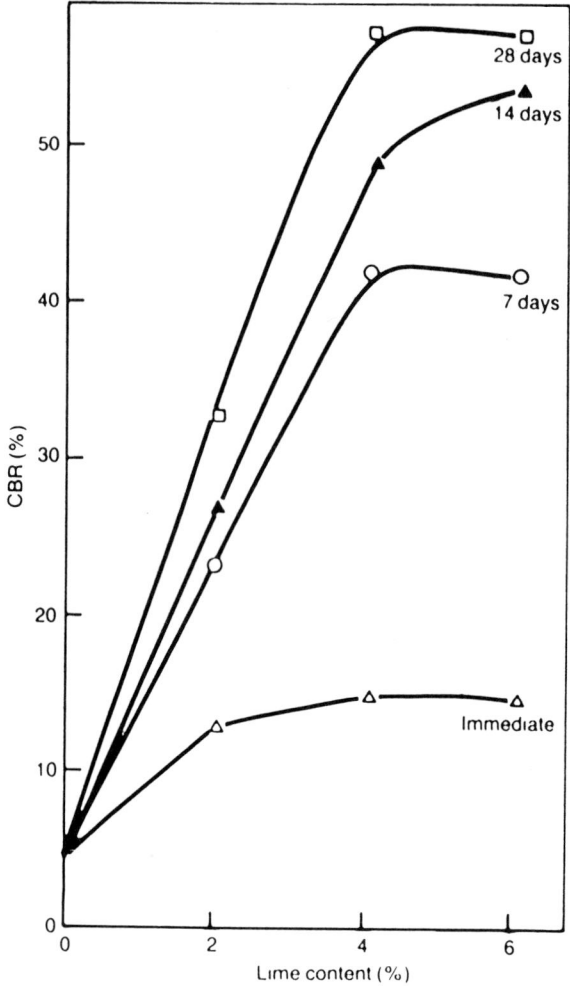

Figure 2   **Gain in strength of lime stabilised clay (after Littleton *et al* 1988)**

Construction of Lime or Lime Plus Cement Stabilised Cohesive Soils

J.H.SMITH
Consultant to Powerbetter Developments Ltd

INTRODUCTION

Although lime stabilisation of cohesive soils has been carried out in the UK since the 1950's, it did not form part of a recognised British Specification until 1986, when it was included in the Department of Transport's Manual of Contract Documents for Highway Works Volume 1 under Clause 615 'Lime Stabilised Capping' (MCHW1 Clause 615). This specification has remained the only recognised method of constructing lime stabilised soils, although other specifications have been developed by the stabilisation industry based on Clause 615 for both bulkfill and sub-base applications.

Lime treated soils for bulkfill have been adopted on a number of notable road and building projects in recent times, including the A13 lime-pfa treated silt reported in this publication by Nettleton et al (1996), where it was deemed necessary to utilise lime stabilisation methods for capping to achieve the necessary degree of mixing, consistency of end product and efficient use of lime.

Although relaxations are generally allowed for bulkfill treatment on end product strength, maturing period and degree of pulverisation, the application of lime using spreaders and the mixing/moisture adjustments using rotovators have been found to be important.

With regard to the higher strength sub-base applications, it is generally found necessary to add cement as well as lime to achieve the required strength and durability.

The construction methods and equipment adopted for lime treated soils are discussed below, including the measures necessary for the health and safety of personnel.

SPECIFICATION REQUIREMENTS

Prior to any soil stabilisation taking place on cohesive soils, a thorough investigation of the candidate soils must be carried out to determine soil suitability and lime or lime plus cement quantities required to achieve the end product specification.

The soil stabilisation specification, which will generally be in accordance with MCHW1 Clause 615, but may have some modifications depending on the end product requirements and the binders to be used, will include requirements for both untreated and treated soils, generally as follows:

Untreated soil

1. Grading: Lime only   100% passing 75mm sieve
                    (as class 7E)   95%-100% passing 28mm sieve
                                           15%-100% passing 63 micron sieve

2.	MCV or m.c.	To enable the stabilisation process to proceed. For cut situations, in general the MCV needs to be at least 6. For filling on a firm layer below, no MCV limit is necessary. As water can be added to a dry soil, no upper MCV limits are necessary.
3.	Plasticity Index:	At least 10% for lime only treatment. No lower limit is necessary for lime plus cement treatment, as there is no reliance on clay content to react with the lime for cementing, given the inclusion of Portland cement.
4.	Organic Content:	Generally no more than 2%, although this may be exceeded if strength and durability can be satisfied.
5.	Total Sulphate and Sulphur Content:	The maximum deemed to be acceptable from prior testing, which ensures that the required soaked CBR is satisfied and acceptable swelling is maintained in the long-term. Generally no more than 1% total sulphate is permitted.

Treated soil

1.	Pulverisation:	95% passing 28mm sieve and 30% passing 5mm sieve for lime only treatment to achieve a capping standard. 60% passing a 5mm sieve is generally adopted when cement is also used. A lower standard may be deemed acceptable for bulkfill applications.
2.	MCV immediately before compaction:	Typically 8 to 12. Here the upper MCV will correspond to optimum moisture content to ensure that 5% maximum air voids is achieved. The lower MCV will ensure that the required soaked CBR is achieved with the level of lime or lime plus cement proposed.
3.	California Bearing Ratio:	Laboratory CBR's after 3 day cure 4 day soak to be 15% for capping. Only the capping strength is currently recognised by the Department of Transport (DTp) as no specifications exist for lime treated bulkfill or sub-base. The Engineer must adopt a suitable CBR for other applications.

Frequency of testing

The frequency of testing recommended by the DTp for road projects given in MCHW2 G100 table NG1/1 is generally adopted for lime stabilisation projects, both in the private and public sector, although some relaxation or increase may be adopted depending on overall project size, variability of soils, soil suitability and actual application involved (i.e. highly suitable and less variable soil for low duty applications may require less testing than other more sensitive schemes).

MCHW2 recommendations for untreated soil are:

Grading/MCV/m.c.	every 400 tonnes
PI	daily
Organics/sulphates/sulphur	daily or twice weekly

MCHW2 recommendations for treated soil are that all tests are carried out once every 500m^2.

CONSTRUCTION PROCEDURE

The construction of lime/cement treated soils - whether for bulkfill, capping or sub-base applications - generally follows the same procedure currently adopted for lime stabilised capping:

1.   Prepare formation to level.

2. Spread quicklime with a purpose built spreader.

3. Undertake initial mixing and pulverisation with a purpose built rotovator.

4. Add water through the hood of the rotovator as necessary to achieve the required MCV.

5. Trim and compact with a grader and roller.

6. Allow to mature.

7. Remix with rotovator and add water to achieve required pulverisation and MCV. Add cement if necessary.

8. Undertake final compaction with roller, checking density and air voids and finish surface with grader within final level tolerances.

9. Cure and protect.

Each stage of this process is considered in detail below, including a review of equipment currently utilised.

Preparation of formation

In-situ stabilisation is generally carried out in the material's final location, either on a cut formation or on material deposited for fill. In this instance, as all the processing is carried out in one area, quality is easier to control and there is less risk of material falling outside the specification limits.

If necessary, material can be treated first and then excavated and compacted in a fill area, but in this instance material must not be allowed to dry out whilst being transported and care must be taken when excavating not to overdig into untreated soil below.

Prior to stabilisation, the formation should be trimmed and compacted to the level of the final treated layer within a normal formation tolerance of +20 -30mm with allowance made for up to 10% bulking, due to the effect of the lime, on the maximum dry density of a cohesive material.

In preparing the formation, it is important to remember that the level and density of the soil prior to treatment will determine the accuracy of the mixing depth, final thickness and overall strength and consistency of the layer.

To reduce the potential for standing water on the stabilised layer which may soften the top surface, the formation should be shaped to falls where possible, and sufficient drainage, either temporary or permanent, should be provided to remove surface water once stabilisation is complete.

In road cuttings where side drainage has not yet been constructed, it is normal to provide temporary grips to collect water from cutting sides.

To minimise the disturbance of the stabilised soil, it is normal to conduct all main drainage prior to soil treatment. Minor gully connections and ducts may still be constructed afterwards.

Before any mixing can take place, all cobbles and boulders above 125mm must be removed either by hand or by excavator bucket, so as not to damage the rotovator and to satisfy the grading specification.

Lime spreading

For accurate spreading and consistent soil treatment, it is necessary to utilise a purpose built mechanical spreader for applying quicklime and cement to the prepared formation surface, which is required by the DTp for stabilised capping. Spreaders are either motorised or towable (see Figures 1 and 2) and are filled in most circumstances from a bulk air tanker (Figure 3), although silo storage may be used on large projects to guarantee lime supply.

The lime dosage is based on the available lime content of the product used, either quicklime or hydrated lime complying with BS890. The required weight of lime is calculated on the dry density of the untreated soil, according to the minimum dosage required after prior testing to achieve the end product. The MCHW1 requires 2% minimum lime content for capping.

Work is generally planned in parallel bays which can be completed within a reasonable time, particularly so that if heavy rain or winds are imminent, lime is not left on the ground for too long to be exposed to adverse conditions.

The spreader discharge systems are designed to spread lime at a constant rate whatever the travel speed of the spreader, which can vary due to differing site conditions and obstructions. The rate of the discharge is adjusted to suit the required dosage. The spread rate is checked by spreading over a tray generally every $500m^2$, which is weighed and the discharge adjusted as necessary to maintain the correct rate. To minimise dusting, skirts are fitted around the rear of the spreaders to reduce the possibility of wind picking up dust as it falls from the discharge slots (see Figure 4).

Initial mixing and pulverisation

Once the lime has been spread on the formation, mixing is carried out by a powerful purpose built rotovator which has a single cylindrical rotor with many tines or cutting teeth. The rotovator is lowered into the ground by hydraulic rams to the required depths to consistently mix the lime with the soil. The rotor is enclosed within a hood which incorporates a spray bar for water addition. More than one pass of the rotovator is usually required to achieve the consistent and homogenous mix necessary during the initial mixing stage prior to maturing. (The MCHW1 requires 2 passes minimum for stabilised capping.)

Although some rotovators are capable of mixing to 500mm depth, the thickness of mixing should be limited to 250mm if MCHW1 method compaction is to be used, although adequate compaction can still be achieved with mixing depths up to 350mm with present day compaction equipment. For this reason mixing may be undertaken up to 350mm and end product compaction satisfied.

To avoid leaving untreated material, adjacent strips are overlapped generally by 150mm and for filling and stabilisation in layers, each layer is cut in by 20mm.

Rotovators There are a number of purpose built rotovators available world-wide to undertake soil treatment with lime, cement or other suitable binders. All are powerful and highly sophisticated items of plant, some with greater strengths than others, but all possessing the necessary features for most soil stabilisation tasks.

Some of the soil stabilisation rotovators currently available in the UK include the Caterpillar SM-350 (Figure 5), Wirtgen Recycler WR2500 (Figure 6) and Bomag MPH120 (Figure 7). Although capabilities vary between machines, some being more suited to certain situations than others, the features are similar in many respects and can be summarised as follows:

1.   A powerful diesel engine, with power output up to 600HP for the larger machines.

2.   Single rotor with depth control (Figure 8).

3.   Liquid Additive System. Water for lime/cement treatment is pumped through
       a bowser and added through a spray bar with individual nozzles into the hood
       of the mixer. Once the water addition rate has been set by the operator, the flow is
       automatically adjusted to account for the speed of the machine for accurate application
       and consistent moisture content adjustment.

4.   Effective Traction. Some machines have a 4 wheel drive facility to ensure
       permanent traction, even when working at great depths or on difficult ground.

The typical width of the rotor is 2450mm. Mixing depths up to 500mm are achievable with the
larger machines. With some rotovators, once the depth is set by the operator, a computerised
monitoring system will automatically adjust the rotor depth. Depth control is critical to ensure
the correct thickness of end product and correct proportioning of lime content.

The tools attached to the rotor vary in size, shape and number. Spade type tools are generally
adopted for soil stabilisation, with L shaped chopper tools for heavy clays. The number of
tools can vary between rotovators from 60 per rotor up to 230 for the more capable and
ultimately expensive machines.

Water addition

During initial mixing and pulverisation, it may be necessary to add water through the hood of
the rotovator from a bowser travelling in tandem with the machine (Figure 9), to initially
hydrate or slake the quicklime (if used) which is then carried in solution to react with the clay.
Water is also added to achieve the required moisture content or MCV range to satisfy
compaction and strength criteria, typically an MCV of 8 to 12.

Potable or concrete standard water should be used, unless the chemical constituents of other
sources can be shown to be acceptable. Where a considerable amount of water is required, for
example with a desiccated clay, spraying directly onto the surface from a bowser rather than
through the rotovator hood may be considered acceptable as a preconditioning exercise, prior to
lime spreading and initial mixing. Final water adjustment through the rotovator hood is
preferred, as greater control and consistency of the mix is achieved. Modern rotovators have
built-in metering systems which enable the operator to make correct adjustments to the water
flow. The spray nozzles under the rotovator hood each have their own tap, so that flow can be
cut off where required, typically where bays overlap, so as not to overwater in such situations.

Trimming and light compaction

Following each mixing pass the material is trimmed and lightly compacted by a smooth
wheeled roller (Figure 10) or pneumatic tyred roller to ensure correct processing depth
throughout the operation. Once initial mixing and water addition is complete, the material is
given a final trim and compaction prior to a maturing period of between 24 and 72 hours for
capping (see below).

Light compaction is required for a number of reasons at this stage:

1.   It brings the lime into intimate contact with the clay.

2.   It minimises evaporation loss.

3.   It reduces possible damage from rain.

4.   It reduces the risk of lime carbonation.

Carbonation occurs when the lime reacts with carbon dioxide in the air and reverts back to
calcium carbonate (limestone) on long time exposure, before it has had time to react with the

clay. If this were to occur, it would obviously reduce the amount of lime available to improve the soil.

Maturing period

For lime stabilised capping, a maturing period of 24 to 72 hours is required to allow modification of the soil to occur (i.e. full change in plasticity, optimum moisture content and MCV). This maturing period allows the lime to slake, provides time for the lime to migrate through the soil, and as a result of the plasticity changes, makes mixing and breakdown of the clay to achieve the required pulverisation much easier. For some applications, such as temporary works or bulkfill treatment, this stage may be omitted and a less stringent pulverisation specified. If less than 24 hours is allowed for the lime to react with the soil, then it is important that the prior testing models the actual maturing period adopted in the field, to ensure that the appropriate MCV and m.c. limits are adopted for adequate compaction, as they can change with time during the maturing/modification stage.

Re-mixing and final moisture adjustment

Following maturing, re-mixing and final water adjustment takes place using the rotovator to finally achieve the required pulverisation and MCV range. The MCV range in the field must ensure that 5% air voids is achieved after final compaction, normally equivalent to optimum moisture contents and an MCV of 12 to 13, and also the required CBR is achieved at the lower MCV, generally around an MCV of 8 to 9.

Final compaction and surface finishing

Adequate final compaction is essential to ensure the long term strength and durability of the lime stabilised layer. MCHW1 details the number of passes of specific rollers for layer thickness of 150 and 250mm in table 6/4 method 7. For alternative thicknesses and/or compaction equipment, a compaction trial is necessary to determine the number of passes to achieve the required density and air voids through the full depth of the layer. For thicker layers, good results are achieved using a tamping sheeps foot, roller (Figure 11) followed by a smooth wheeled roller. During this final compaction stage, surface irregularities can be corrected by trimming using a grader.

Curing and protection

As with other cementitious materials, adequate curing is essential for maximum development of strength and durability. It is important therefore that stabilisation is carried out in favourable temperatures and the moisture content of the material is maintained at a satisfactory level. MCHW1 requires a minimum temperature of 7°C for lime stabilised capping, with work only carried out during March to September, unless agreed otherwise by the Engineer. It has become normal practise however to work outside the March to September period provided temperatures remain acceptable around the 7°C level.

With regard to prevention of surface drying, the following measures can be adopted:

1. Placing the next layer without delay.

2. Frequent spraying of the surface with water.

3. Adopting a bitumen seal coat.

Light rain during stabilisation need not stop the work, although lime will not be spread during periods of rain. In the event of heavy rainfall, the bay under construction will be completed as quickly as possible and work suspended.

Traffic may be allowed on the treated layer, preferably after 2 to 3 days curing, provided it is not rutted or distorted by the loading. It should be noted that a capping material is only designed to perform the function of a working platform for the laying of the sub-base. It is not designed to carry normal construction traffic which is the function of the sub-base, and will inevitably be distressed after a short time without additional protection, particularly during wet weather.

HEALTH AND SAFETY

Risks associated with quicklime

Apart from the normal risks associated with the operation of soil stabilisation equipment - which will require appropriate safety measures to be implemented by site personnel - the single element requiring particular attention to ensure the health and safety of both operatives and the public, is the handling and mixing of quicklime.

When quicklime (calcium oxide) comes into contact with water, be it water in the soil or perspiration on the body, it will hydrate to form hydrated lime (calcium hydroxide) and in doing so will release considerable heat with the potential to cause burns. The resulting hydrated lime is a caustic alkali, which in turn may also cause chemical burns. Quicklime is classed as an irritant to eyes and skin, the main risk being that of impaired vision or blindness due to the potential for burning. It can also cause internal damage if inhaled or ingested and this too must be avoided. Adequate personal protection to guard against the risks outlined should be provided for personnel likely to come into contact with quicklime (see below for specific protective equipment). Obviously those personnel requiring the greater degree of protection will be those operating the spreading and mixing equipment. Personnel not actively involved in the lime stabilisation operations should wear protective equipment also to guard against the effects of windborne lime dust. No one should be allowed to approach an area whilst lime spreading and mixing is in operation without suitable eye protection. Eye wash facilities at clearly marked stations should be provided in case of emergencies.

Protective equipment

Eye Protection Goggles must be worn to prevent lime dust entering the eye. Wide vision, full goggles with anti-mist properties are preferred or an air stream helmet. Safety spectacles are less suitable.

Respiratory Protection An approved dust respirator should be worn as required. A dust mask consisting of gauze-covered cotton wool filter pads held in a wire frame with a headband is effective for the mouth and nose.

Protective Clothing Long sleeve shirt and trousers should be worn and not short sleeve or rolled up sleeves or shorts. Gloves should also be worn, preferably with a knitted wristband. Suitable boots to prevent lime reaching the feet should be worn, with gaiters as additional protection if necessary.

Exposed Skin Any exposed parts of the body, particularly those where perspiration is excessive or skin is sensitive such as shaven parts, may be protected by a barrier screen.

Precautions to guard against dust

As well as providing the necessary protective equipment, there are a number of precautions which should be adopted in the handling of quicklime to reduce the atmosphere levels of lime dust:

1. Spreading of quicklime should not be undertaken in strong winds, particularly when there is a risk of lime blowing off site.

2.   Plant should not be allowed to drive over spread quicklime, which has not yet been rotovated into the soil.

3.   Additional precautions in sensitive locations adjacent to housing, schools, etc., should be taken. For example, when working next to schools, it maybe necessary to limit spreading when pupils are not in the playground.

4.   Mixing should follow on close behind the spreading to minimise the time that lime remains on the ground to be picked up by the wind.

First aid

In the event of First Aid being necessary, the typical procedures recommended by lime producers (e.g. BLI, undated) in their Chemical Data Sheets - issued under the Health and Safety Commission Chemicals (Hazard Information and Packaging) Regulations to all purchasers of quicklime - are reproduced below.

Skin	Remove contaminated clothing. After contact with the skin, wash immediately with plenty of water.
Eyes	SPEED IS ESSENTIAL. Particles should be removed with a cotton wool bud. Irrigate with eye wash solution or clean water for at least 10 minutes. Obtain immediate medical attention. Continue irrigation until medical attention can be obtained.
Inhalation	Remove patient from exposure, keep warm and at rest. The nose and throat should be thoroughly irrigated with water for at least 20 minutes.
Ingestion	Do not induce vomiting. Wash out mouth with water and give copious quantities of water to drink.
Further Medical Advice	Symptomatic treatment and supportive therapy as indicated.

REFERENCES

British Lime Association (1990), "Lime Stabilisation Manual", BLA.

Buxton Lime Industries (undated), "Chemical Data Sheet for Quicklime", BLI, Buxton.

Manual of Contract Documents for Highway Works, Volume 1: Specification for Highway Works (December 1991 with amendments), HMSO, London.

Nettleton, A, Robertson, I and Smith, J H (1996), "Treatment of Silt using Lime and PFA to form Embankment Fill for the New A13", Lime Stabilisation, Edited by C D F Rogers, S Glendinning and N Dixon, Thomas Telford Ltd, London.

Figure 1   Spreader used to distribute lime

Figure 2   Spreader used to distribute lime

Figure 3   **Bulk air tanker used to fill spreader**

Figure 4   **Skirts fitted to spreaders to reduce dust formation**

Figure 5   Caterpillar SM-350 rotovator

Figure 6   Wirtgen recycler WR2500 rotovator

Figure 7   Bomag MPH120 rotovator

Figure 8   Detail of rotovator showing single rotor with depth control

Figure 9   Addition of water to the hood of a rotovator during mixing

Figure 10   Compaction of a lime stabilised clay using a smooth
wheeled roller

Figure 11   Compaction of a lime stabilised clay using a tamping sheeps foot roller

The Uses of Lime in Ground Engineering: a review of work undertaken at the Transport Research Laboratory

J PERRY
Transport Research Laboratory, Crowthorne
D J MACNEIL
Transport Research Laboratory, Crowthorne
P E WILSON
Highways Agency, London

INTRODUCTION

This Paper considers recently completed and on-going research by the Transport Research Laboratory (TRL) into the uses of lime in ground engineering. The applications presented are:

1.   highway capping;
2.   bulk fill;
3.   railway track bed;
4.   slope stability;
5.   contaminated materials.

The benefits to the Client are outlined before presenting the technical support for the use of lime. Guidance is given on design, including site investigation strategy and laboratory testing, and application. The main critical soil properties are discussed and their detection and evaluation briefly described. In addition, consideration is given to the chemical and geotechnical effects stabilisation may have on cohesive soil. The use of quicklime only is the principal concern of this Paper; hydrated lime, cement and other additives are also being considered as part of the research programme at TRL.

Some clarification of terms is first required, however, and this is given below.

Lime modification

When lime is mixed with cohesive material, the material is first modified before further chemical reactions occur which in most soils lead to stabilisation. Not all cohesive materials will stabilise but all will modify. Mixing quicklime with a wet soil immediately causes the lime to hydrate and an exothermic reaction occurs, the heat produced being sufficient to drive off some of the moisture within the soil as vapour and hence reduces the moisture content. The second effect of ion substitution results in a reduction in plasticity as the clay particles flocculate. Modification with lime results in considerable changes in the engineering characteristics of a treated soil, and has been used as a mitigation measure on a number of waterlogged sites (Sherwood, 1992).

Lime stabilisation

After lime modification, a period can be allowed of between 1 hour and 72 hours, depending on application and time available, to allow the material to mellow. This period allows the quicklime to slake, provides time for the lime to migrate through the soil and, as a result of the plasticity changes, makes mixing of the material before final compaction easier. In the longer term the lime reacts with the clay particles to produce cementitious products which then bind the soil together.

Both modified and stabilised cohesive materials will require final compaction and considerable amounts of water may need to be added as the soil is mixed. It is essential that the moisture content is greater than the optimum moisture content otherwise inadequate compaction will occur.

Sulphur

Sulphates, for example gypsum, react with lime and cause swelling which can be detrimental to the material's strength and cause deformations of any final surface (Sherwood, 1993). The sulphates may be present within the soil already, be produced by the oxidation of sulphides, or be introduced by groundwater. The state of compaction has a significant effect on the amount of heave associated with sulphate and sulphide reactions. Adequate compaction, that is 5% air voids or less for lime treated bearing layers or 10% air voids or less for modified bulk fill, can help to reduce heave in both the short term, during the construction period, and long term. Organic materials may also prevent the stabilisation process occurring, the effect being dependent on the type and amount of organic materials (Sherwood, 1993).

Curing Period

The time after final compaction.

Mellowing Period

The time between mixing a soil with lime and final compaction.

Swell

The linear expansion of a sample in a CBR mould when subjected to soaking. The value is either expressed as a percentage of the height of the CBR mould or as an absolute value.

CLIENT'S AIMS AND OBJECTIVES

In the following section, the aims and objectives of the Highways Agency are given as an example of the benefits to be gained from using lime. These benefits are given to add perspective to the use of lime and can also be relevant to other clients who are seeking similar goals.

The Secretary of State's aim is that the Highways Agency should secure the delivery of an efficient, reliable, safe and environmentally acceptable trunk road network (Highways Agency, 1996). In order to meet this aim, the Highways Agency is committed to four key principles. These principles underpin everything that is done to manage, maintain and improve the trunk road network within the Government's legislative policy and resource framework.

The four principles are outlined below together with a brief indication of how lime can help the achievement of the client's aims and objectives.

1. *Serving the public* by identifying and balancing customer needs and agreeing service levels with customers and programmes with Ministers. The primary customers are the users of the trunk road network but there are also responsibilities to those affected by the trunk road network. The use of lime can assist in reducing the impact of construction on local communities by reducing the volume of construction traffic using local roads. The A13 Thames Avenue to Wennington is a good example of this (Nettleton *et al*, 1996).

2. *Serving the road user* by delivering services to agreed standards of quality, time and cost; and by developing a staff who are proficient and valued, and a quality culture concentrating on performance and encouraging innovation. Reliability of journey time on the network is a key concern to road users. The management and maintenance of the existing network with minimum delay to the road user is increasingly important. The occurrence of shallow slips on the ageing highway earthwork slopes are predicted to increase (Perry, 1989). Lime offers a variety of techniques to repair shallow slips without causing delay to the road user. The use of lime in embankment repairs (Johnson, 1985) and for stabilising slopes with lime piles (Rogers and Glendinning, 1993) are potentially useful ways of reducing delays to the road user.

3. *Protecting the environment* by delivering services and measures to minimise the impact of roads on the environment. The Highways Agency is committed to striking a balance between the environmental and economic costs and benefits. Lime stabilised soils can be used as a capping layer as an alternative to imported granular materials; this reduces the demand for quarrying natural resources (HA 74 (DMRB 4.1.6); Sherwood, 1993).

4. *Serving the taxpayer* by improving value for money. The Highways Agency will continue to improve risk identification techniques and cost estimating techniques. The main source of cost increases and delays in highway construction contracts are unforeseen ground conditions. Lime modification of wet cohesive clay fills enables unacceptable fill material to be rendered acceptable. This avoids delay and disruption to the contractor's work programme and allows maximum use of on site materials (National Audit Office, 1992; Mott MacDonald and Soil Mechanics Ltd, 1994).

Other clients have similar aims and objectives as outlined above but directed toward other areas. For example, in railway maintenance there is a great need to minimise costs and delays to train movements and to improve the existing rail network. Lime modification to improve track bed and the use of lime piles on unstable slopes both assist in solving these problems. New railway construction is also an area where maximum use of material can be effected by using lime.

The following sections of this Paper describe the programme and results of research currently underway or recently completed at TRL which has the clear direction of meeting these aims and objectives.

DESIGNING LIME STABILISED HIGHWAY CAPPING

Capping is a higher strength layer placed and compacted on weak fills and cutting foundations. Material used in the capping must be of sufficient strength to provide a working platform for construction of the pavement layers and act as a structural layer in the longer term. This can be achieved by using either granular materials, of fine or coarse grading, or materials stabilised with lime or cement to the requirements of MCHW 1. In order to determine the suitability of a site for lime stabilisation, a site investigation, which considers the use of lime as an option, is needed to ensure the design is feasible.

Site investigation (SI)

The preliminary sources study (PSS) stage of the SI should ensure that lime stabilisation of capping is considered as an option from the very beginning of the SI and earthworks procedures. The PSS should include an assessment of local geology and geotechnical features associated with the route.

The purpose of the ground investigation (GI) is to provide information on which to judge, after consideration of the PSS, whether lime stabilisation is viable and to provide sufficient material to carry out testing to ascertain an adequate mix design. The strategy adopted to meet this purpose should be geared toward obtaining the maximum amount of representative data for the minimum cost and in a reasonable period of time. Advice on both GI strategy and PSS methodology are detailed in Perry *et al* (1996) and HA 74 (DMRB 4.1.6).

Limits on the properties required for cohesive material before adding lime, and the limits once lime has been added, are essential to allow both the selection of suitable material and to give assurance of good long term performance of the pavement structure. Suggested laboratory tests to determine these limits for design purposes are given in Table 1 and a suggested test procedure is shown in Figure 1. Tests for suitability are those to determine whether or not the material is able to be stabilised. Tests for acceptability are for determining whether the soil meets the requirements of the Classes in MCHW 1 or for providing values for inclusion as limits in the highway contract.

Additional advice is given in Perry *et al* (1996) with respect to soil tests for suitability and acceptability of material for stabilisation (Class 7E), and for stabilised material (Class 9D). This advice encompasses Plasticity Index, Moisture Condition Value (MCV) and moisture content, sulphates and total sulphur content, organic matter, Initial Consumption of Lime, California Bearing Ratio (CBR) and swell, lime addition, laboratory compaction and frost susceptibility. Due to limited space, full details of all these tests cannot be encompassed in this text. However, as the MCV test has demonstrated considerable applicability to both lime treated and untreated cohesive soils, a brief outline of the application of the MCV test follows.

For Class 7E material the limits on MCV, or moisture content, only need to reflect the limits on earthworking plant operation and material handling and hence only a lower limit is required. The value would typically be around an MCV of 7, although some stabilisation plant can cope with sites wetter than this. The lower and upper limiting values for the Class 9D material are based upon compaction requirements and the laboratory CBR. The lower limit would again be about an MCV of 7: the upper limit would be dependent on ensuring that the material had sufficient moisture to achieve a specific air void target of 5% air voids or less.

Lime stabilised soils behave in a very different manner to most naturally occurring soils and have completely different plasticity characteristics to the original unstabilised material; research has shown that the MCV is able to reflect these changes. MCV is less dependent on time of testing than optimum moisture content. Figure 2 illustrates this point for a stabilised heavy clay. In the figure, it can be seen that the optimum moisture content has changed as the stabilised heavy clay becomes more granular in behaviour with increasing time but the MCV at the optimum moisture content still stays around 13.5 (the calibration leg which must be used is the 'wet leg', which shows reducing MCVs with increasing moisture content values). Consequently a specified limit of an MCV of 12.5 for a stabilised heavy clay would be wet of optimum moisture content and would remain so independent of time. Figure 3 (based on Parsons, 1992) shows that the MCV at optimum moisture content varies between about 12 and 14 for a number of stabilised materials ranging from lime stabilised glacial till and heavy clay to gravel-sand-clay. The MCVs for the different materials given in the figure are considered an absolute maximum for lime stabilisation and it is essential for compaction purposes to be at lower MCVs, that is to be 'wet' of optimum moisture content. The relation between MCV and optimum moisture content is the key to why the MCV is such a useful test for lime stabilisation; although the plasticity properties of a Class 9D material vary with time, the MCV remains relatively constant.

Conclusions

It is recommended that the site investigation should:

1.   include the early identification of sulphides and sulphates in the field;

2.   include laboratory testing for sulphides and sulphates, if not found in sufficient amounts in the field to make lime stabilisation unfeasible;
3.   emphasise achieving adequate compaction: the MCV is the best means of doing this;
4.   be guided by an upper MCV of between 12 and 14 to ensure compaction wet of the optimum moisture content for all materials stabilised with lime;
5.   control swelling of the capping by:
        (a)   putting limits on swelling during extreme laboratory soaking tests;
        (b)   advising that the total sulphate content should be below 1%;
        (c)   investigating the potential production of sulphates from sulphides;
        (d)   providing compaction to 5% or less air voids;
6.   use organic matter limits based on laboratory test results for Initial Consumption of Lime and California Bearing Ratio;
7.   emphasise that tests should be conducted at the same time and that these times are representative of site requirements. This leads to a contradiction with the test in British Standard BS 1924: 1990 for the optimum moisture content of stabilised material.

MODIFICATION OF BULK FILL

Due to concerns regarding the costs associated with the rejection of wet cohesive fills, a research project was undertaken to assess the viability of rendering unacceptable materials as acceptable by the addition of lime. For lime modification of wet cohesive soils, the basic requirement is the rapid amelioration of the soil for acceptable shear strength, trafficking and compaction purposes, and the strength gains associated with the stabilisation mechanism are not required. In the longer term, the more demanding requirement is to ensure durability of any earthwork constructed using lime modified soils.

Test methodology

Two soils, representative of the upper and lower ends of the plasticity ranges encountered with typical British cohesive soils (characteristics detailed in Figure 4 and Table 2 and classified as Class 2A in MCHW 1), were modified with various additions of SG60 quicklime. The heavy clay was treated with additions of 0.5%, 1.0% and 2.5% (the minimum addition recommended in MCHW 1 for lime stabilisation to form capping) and the sandy clay was treated with 0.5% and 2.5% additions. The methods of test specified in BS1377 (BSI, 1990a) and BS 1924 (BSI, 1990b) were followed. After mellowing periods of between 4 and 48 hours, and following the advice in HA 74 (DMRB 4.1.6), the treated materials were tested for laboratory compaction (2.5kg rammer), particle density, Moisture Condition Value (MCV) calibrations, immediate California Bearing Ratio (CBR), 7 day soaked CBR, long term (nominally 28 days) soaked CBR and swell as a result of immersion in water. The shortest mellowing period of 4 hours was selected as a compromise between the typical mellowing period used so far on site and the best estimate of the time after which most of the pronounced changes in the material would have occurred.

Results

The laboratory compaction data for the heavy clay (Figure 5) display the typical non-classical profiles associated with lime treated cohesive soils. With increasing lime additions, the treated materials tend to produce lower densities, and also require higher moisture contents to achieve specific air void targets. For the treated heavy clay, the shift in the MCV-moisture content calibrations (Figure 6) is considerable and reflects significant changes in engineering properties.

The MCV calibrations for the 0.5% and 1.0% lime additions displayed very little movement versus time. In contrast, the MCV calibrations for the 2.5% lime additions do display time dependent behaviour, indicating that stabilisation had occurred. For the 2 day mellowing period, the

Table 1 Soil tests for suitability and acceptability

Material Property	Defined and tested in accordance with:	For suitability	For acceptability
Plastic limit	BS 1377: Part 2	x	
Liquid limit	BS 1377: Part 2	x	
Plasticity index	BS 1377: Part 2	x	
Particle size distribution	BS 1377: Part 2	x	
Organic matter	BS 1377: Part 3	x	x
Total sulphate content	BS 1377: Part 3	x	x
Total sulphur content	BS 1047	x	x
Initial consumption of lime	BS 1924: Part 2	x	
CBR	BS 1924: Part 2	x	x
Swelling	BS 1924: Part 2	x	
MCV for Class 7E	BS 1377: Part 4[1]		x
MCV for Class 9D	BS 1924: Part 2		x
Optimum moisture content for Class 9D (2.5 kg test)	BS 1924: Part 2		x
Frost susceptibility	BS 812: Part 124	x	

[1] In Scotland, DMRB 4.1.4 SH7/83

Table 2 Analysis of cohesive soils

Soil	Sulphate and organic matter analysis[1]				ICL[2]	X-ray diffraction analysis			
	Total sulphate (% SO$_3$)	Water soluble sulphate (g/l SO$_3$)	pH	Organic matter (%)	ICL value (%)	Illite (%)	Illite-smectite (%)	Kaolinite (%)	Chlorite (%)
Heavy clay	0.98	1.62	7.6	1.0	1.9	29	57	14	-
Sandy clay	< 0.01	-	8.3	0.8	1.5	26	61	11	2

[1] BS1377 (British Standards Institution, 1975)
[2] Initial Consumption of Lime, BS1924 (British Standards Institution, 1990b)

Table 3 Moisture contents required to produce material at wet end acceptability range

Soil	Lime addition (%)	Moisture content (%)	Extension of moisture content maxima (%)
Heavy Clay	0	33	-
	0.5	43	10
	1.0	50	17
Sandy Clay	0	19.5	-
	0.5	24.5	5

calibration shows the beginning of a 'dry leg'. The less cohesive sandy clay also displayed similar trends in MCV calibrations.

For heavy clay at a moisture content of 35%, equivalent to an MCV of 6.5, lime treatment was found to typically reduce moisture content by an amount equivalent to the lime addition although the reduction in moisture content increases for moisture contents in excess of 45%. For such a material at a moisture content of 35%, treatment with a 0.5% quicklime addition results in an MCV of typically 12.0. Similarly, treatment with a 1.0% addition results in an MCV of 13.0. Treatment with a 2.5% addition results in MCVs more dependent on time and ranging from 14 to 16.

The effectiveness of lime modification in extending the cohesive soil moisture content maxima, typically equivalent to an MCV of 8, is summarised in Table 3. As a typical moisture content range for the majority of heavy clays found in Britain is 23% to 38% (Parsons, 1992), it is probable that a lime addition of 0.5% would suffice for the treatment of most British heavy clays which have been classified as unacceptable as a result of excess water.

The CBR data for the heavy clay (Table 4) reveal the considerable increases in bearing capacity of the modified materials compared to the untreated soil. The long term CBR data indicates the durability of the lime modified materials. It is also important to note the higher strength of the heavy clay at shorter mellowing periods. The swell data for the 0.5% and 1.0% lime additions, generated whilst the CBR specimens were soaking, were within the swell range of 11mm to 2mm obtained for the untreated material at moisture contents in the acceptability range of MCV 12 to 8 (26% to 33% moisture content). The time taken to achieve steady swell states for the lime modified specimens were of a similar magnitude to, or less than, those for the untreated soil.

Table 4 Typical strength improvement for treated heavy clay

Moisture content (%)	Quicklime addition (%)	Mellowing period (hours)	Imm[1] CBR (%)	7 days[2] CBR (%)	LT[3] CBR (%)
35	0	-	1	1.5	1.5
	0.5	4	4.5	5	5.5
		24	4	4	4
		48	4	-	4.5
	1.0	4	12	11	11
		24	9.5	-	-
		48	9.5	-	-
	2.5	4	19	26	35
		24	18	-	29
		48	19	-	24

[1] Immediate CBR
[2] 3 days dry/4 days wet cure
[3] Long term, 3 days dry/25+ days wet cure

Conclusions

Considerable benefits can be realised by using lime modification to treat wet cohesive soils. Of great significance is the fact that additions of lime which produce lime modification of cohesive soils do not significantly alter the moisture content of the modified soils compared to the untreated materials. However, the lime modification process does significantly alter engineering characteristics, as indicated by changes in compaction curves, MCV calibrations, bearing capacities and Plastic Limits. Using lime modification, the moisture content maxima for wet cohesive soils, which can feasibly be modified with lime and then used as general fill materials, may be extended by considerable margins. The moisture content limits for the untreated material should no longer be used for the modified wet cohesive material as plasticity changes have also occurred.

With respect to advice on appropriate testing for lime modified materials, it is recommended that the methods of test for lime stabilisation, with the exception of meeting strength (CBR) requirements, be followed. The swell requirement for lime stabilisation could also be replaced by a more appropriate requirement, such as the range of swells associated with the untreated materials when compacted across their range of acceptable MCVs.

It would also appear from the laboratory results that a higher CBR value can be achieved with little or no mellowing. This is due to the improved cementing that can occur if the material is compacted immediately after mixing with lime rather than breaking any bonds that had formed during a mellowing period. However, on-site materials which are excessively wet may need some mellowing period to allow the diffusion of lime, improved trafficability and mixing, and since the improvement in strength is still very significant, immediate compaction may not be advisable.

TREATMENT OF RAILWAY TRACK BED

In 1995, TRL was commissioned by Railtrack to consider the application of the addition of lime to existing rail track bed for situations where very little time was available during track possessions. Railtrack considered that the use of lime could offer considerable benefits for the improvement of clay, London Clay in particular, track bed performance by increasing undrained shear strength and modulus, and by decreasing the susceptibility to subgrade erosion. The aim of this study was to ascertain if the bearing capacity required could be achieved within this time scale. Consideration was also to be given to the use of thick layers treated with lime as a means of reducing ground vibration from trains.

The time available for adding and mixing lime during a track possession is only a few hours and so the mellowing period will be much shorter than is used during construction. This short time-scale also means cementitious gels will not have time to fully develop before being subjected to track loading. As such, the design of the lime treated soil is based on a lime modification effect with strength increases based on the short-term gains rather than long-term pozzolanic formations. However, in the longer term, further gains in strength will inevitably occur after the track is back in use.

Test methodology and procedure

Properties of the heavy clay, which is of London Clay origin, used in the laboratory trial were presented previously in Figure 4 and Table 2. These are typical soil properties for the weathered and unweathered cohesive materials of the London Clay (Cripps and Taylor, 1986).

Using a laboratory testing regime similar to that required for lime stabilisation in HA74 (DMRB 4.1.6), London Clay at a range of moisture contents was mixed with 2.5% of SG60 quicklime. As the ICL value of the London Clay was 1.9%, a lime addition of 2.5% would provide: good on-site distribution of the lime during mixing at an economic rate; long-term strength increases as stabilisation occurs; added confidence in the long-term performance of the track bed. After mellowing periods of 1 hour and 2 hours, the modified materials were re-mixed and tested for compaction characteristics and strength (immediate CBR). Methods of test, as set out in BS 1377 (BSI, 1990a) and BS 1924 (BSI, 1990b), were followed.

The Railtrack strength requirement for track bed is for a subgrade (undrained) modulus of between 80MPa and 130MPa. This criterion was used for acceptability limits for the treated London Clay. In addition, a maximum air void content of 5% was used as the target for adequate compaction; this reflects the higher loading capacity needed of the track bed compared to bulk fill. As both strength and compaction can be related to the Moisture Condition Value (MCV) (BSI, 1990a), the effectiveness of the MCV for such short mellowing periods needed to be ascertained as it provides the best method for material control in lime stabilised soils (HA 74 (DMRB 4.1.6)). The deleterious effect of sulphates was also addressed using the swell tests, however *in situ* it is unlikely that sulphates will be present within the upper 1.5m of track bed as most sulphates have been removed by weathering over the past 150 years or so since construction.

Results

For 2.5% lime addition and 1 hour and 2 hours mellowing period, the compaction curves displayed typical non-classical profiles whereby an optimum moisture content could not be easily identified (similar to those shown in Figure 5). To enable pertinent moisture content (or MCV) limits to be selected so that the modified materials will meet the acceptability criteria for the intended engineering use, the 5% air void intercept was used to define the drier end of the acceptability range. The shift in the MCV calibrations, as a result of modification (Figure 7), was considerable and reflected significant changes in engineering properties. Regression analysis of the MCV/moisture content data indicated that the best fit curves are linear for Class 2A materials and exponential (decay) for the lime modified materials. The difference between 1 hour and 2 hour calibrations was very small.

CBR data for the modified and untreated London Clay was converted to subgrade (undrained) modulus (E) using the approximate formula: $E = 17.6 \, (CBR)^{0.64}$ MPa (HD25 (DMRB 7.2.2)), and are plotted in Figure 8. Comparison between similar moisture content levels for the untreated clay and clay modified with 2.5% lime addition reveals large immediate improvements.

The work undertaken for modification of bulk fill showed that, in the longer term, the CBR will continue to increase for short mellowing periods. Soaked CBR testing was undertaken at 7 days after compaction (3 days dry curing/4 days soaking in water) and at 28 days after compaction (3 days dry curing/25 days soaking) for a 4 hour mellowing period. The data (Table 5) show an increase of strength with time even under soaking conditions. Consequently for the moisture contents at which 5% air voids occurs, i.e. between 30% and 35%, track bed would be expected to continue to gain strength after loading. Swell data (Table 5) was recorded whilst the 28 day CBR specimens were soaking. The amount of swell is directly related to air void content at the time of compaction. The 5% air void target for adequate compaction results in a swell of only 2.56mm (2% of the height of the CBR mould). In the field this swell would be expected to be spread over a long length or be uniform over the length of the modified layer. This amount of swell is well within the tolerances given in HA74 (DMRB 4.1.6).

Table 5 Strength/swell data for 4 hour mellowing of London Clay treated with 2.5% SG60 quicklime

Moisture content target (%)	Immediate CBR (%)		7 day CBR (%)		28 day CBR (%)		Air voids (%)	Swell (mm)	Time to achieve steady state (days)
	top	base	top	base	top	base			
30	20	18	23	28	22	45	8.2	4.28	13
35	17	19	21	26	22	35	5.2	2.56	13
40	12	13	24	30	21	30	2.3	1.31	23
45	7	7	14	25	22	25	0.9	0.76	17

Discussion and conclusions

It has been shown that the improvement, as a consequence of the lime modification process, of weak cohesive soil results in modified materials which could meet the specified strength requirement for track bed.

Based on the results of this research, a specification has been prepared for the treatment of London Clay track bed with lime. The specification also encompasses current knowledge with respect to grading requirements, Plasticity Index lower limit, organic content, an upper sulphate content limit and MCV limits. Three possible approaches to compaction were considered: conducting a trial to achieve 5% or less air voids in a layer and using this compactive effort as a method specification; using 5% or less air voids as an end-product specification for all the works; use the method specification in MCHW 1 for lime stabilised material (Method 7). Due to its successful application in lime stabilisation of capping, compaction to Method 7 was recommended for modified track bed, though the implications of any alteration in terms of layer depth or compactive effort would have to be assessed via a trial. Where track possession time is very limited, the method specification has considerable benefits in terms of speed of construction and reduced testing.

With respect to vibration mitigation, one aspect which requires consideration in determining the effectiveness of the lime modified layer on groundborne vibration would be the long term stiffness of the layer. While this parameter can be easily measured in the laboratory, the effect of vibration on the cements formed during curing is not clear. However, it seems possible that a track bed stiffened by lime stabilisation could be effective in reducing levels of groundborne vibration. In order to confirm this and quantify the reduction, it has been recommended that a site trial be undertaken where measurements are made both before and after the lime treated layer is placed. It would be useful to assess the effects of different layer thicknesses on their use as a vibration mitigation layer.

The results indicate that although a modification process occurs due to the short mellowing period, the railway requirements dictate that testing (using HA74 (DMRB 4.1.6)) and engineering properties to stabilisation standards are needed due to the proximity of the applied rail load.

LIME PILES AND COLUMNS

The Highways Agency requires a simple, economic and effective method of providing stability for the ageing earthworks slopes on the trunk road network (Perry, 1989). The method should be long lasting and not give rise to maintenance problems in the future; it also should not require excessive lorry movements during installation or much disposal of spoil. The use of lime piles and lime-stabilised soil columns are a possible means of doing this. These two techniques are under investigation as part of another TRL project.

Lime piles (small-diameter boreholes filled with quicklime) have been proposed as a method of improving slope stability, and research is currently underway to consider their possible use for this purpose. Lime-stabilised soil columns (larger-diameter holes filled with lime-stabilised soil mixed in place) have been used for many years to improve bearing capacity and reduce settlement and, because of their similarity to lime piles, are included in the research. The proprietary Colmix process, which is a method of producing *in situ* stabilised soil columns using a mixture of lime and cement as the binder, is also being considered in the literature review.

Existing work in this area (Glendinning and Rogers, 1996) has provided a good basis upon which further research can be carried out and has raised many questions as to how the techniques work and to their future integrity. These questions require further work to establish the techniques in the United Kingdom.

The research at TRL, which is in its early stages, will review current methods in the UK and internationally, carry out trial soil and lime mixes, produce and assess different design methods, and consider the long-term durability of the techniques. The review of the available literature on lime piles and lime-stabilised soil columns has recently been completed and will provide valuable information on the success or otherwise of the techniques. It has been shown that all the clays most at risk from slope instability in the UK are likely to respond to treatment with lime. The following conclusions have also been reached.

Lime piles

The beneficial effects of drying of the surrounding clay by quicklime will slowly dissipate with time, so that in the long term the clay will return to its original moisture content and pore water pressure. Available evidence on the migration of lime into the surrounding clay suggests that the lime penetrates only to a small distance. Therefore, the beneficial effect of pozzolanic reaction between lime and clay will be restricted to a skin around the pile. Because of these reasons, ordinary lime piles (i.e. non-reinforced lime piles) would only contribute to slope stability in the long term by virtue of the shear strength of the pile being greater than the shear strength of the ground in which it is installed. Research is needed to establish this quantitatively. Reinforced lime piles (i.e. lime piles containing a steel reinforcing bar down the centre) would contribute more to slope stability than ordinary lime piles because of the additional shear strength of the reinforcing bar. Also, the lime surrounding the bar would perform the useful function of offering it some protection from corrosion. It might be possible to use reinforced lime piles as soil nails carrying some tensile load, subject to overcoming installation difficulties.

Lime-stabilised soil columns

Lime-stabilised soil columns were developed specifically to improve the bearing capacity and reduce the settlement of soft clay. However, because the columns also have a greater shear strength than the ground they are installed in, they could be used to improve slope stability in the same way as proposed for lime piles. To be successfully used in this way, the auger would need to be powerful enough, and robust enough, to drill through, and mix, stiff clays. Lime-stabilised soil columns do not give rise to spoil, in contrast to lime piles which do. Although the volume of material in each hole is small, the large number of holes sometimes required can lead to considerable volumes of soil. This may be an important consideration if lorry movements are to be restricted. Whether spreading the spoil onto the slope is detrimental to stability or is aesthetically undesirable will be a consideration.

Colmix process

The Colmix process offers the advantage of a proven method of installing stabilised soil columns in embankment slopes of in-service earthworks. It has the facility to formulate the lime and cement content of the binder to suit the soil to be treated, and to produce columns that are stronger than if lime alone were used as the binder. It does not give rise to spoil.

Knowledge of the use of lime, site instrumentation, slope instability of man-made slopes, laboratory testing and site trials will be used to determine the suitability of lime piles and lime-stabilised soil columns and the mechanisms involved in improving stability. This substantial and long-term project is aimed at providing guidelines for design and installation to enable the method to be used cost effectively, safely and in a variety of different situations.

PROCESSING OF CONTAMINATED MATERIALS FOR FILL

In addition to the improvement of natural soils, lime may have the potential to be used for the stabilisation of contaminated material. By mixing with lime, often in combination with other inorganic cementitious materials, the contaminants may be encapsulated in cementitious compounds and the physical properties of the material improved such that the material may be reused in construction or more easily disposed of to landfill. (Lime stabilisation in this application, therefore, not only means a long-term increase in strength due to pozzolanic reactions but also the ability of that reaction to encapsulate contaminants within the soil.) Stabilisation is considered an established technology for the treatment of certain inorganic contaminants in the United States, and examples of the successful use of lime in this context are given by MacKay and Emery (1993) and Harris *et al* (1995). The recent use of lime and PFA to stabilise very soft, slightly contaminated silt on the A 13 is discussed by Nettleton *et al* (1996). The purpose of this research is to take this approach further and use lime to stabilise more heavily contaminated material.

The use of lime and other cementitious agents for the stabilisation of contaminated land has several advantages (Perry *et al,* 1996): it gives rapid results, unlike many other methods of treating contaminated land; it can be used for a wide range of soil types and contaminants, particularly inorganic materials; the technology is familiar; the geotechnical properties of the stabilised material can be predicted from standard laboratory tests; construction plant and materials are readily available and the method is relatively cheap; otherwise unacceptable materials can be transformed into acceptable fill, saving import of materials and landfill charges, minimising the volume of construction traffic and disturbance to residents.

Stabilisation is thus environmentally friendly, economically favourable and the practicalities are technically uncomplicated. However, there are some uncertainties regarding the long-term stability of stabilised materials. As the contaminants are not destroyed by the stabilisation process, it is possible that they could leach out if the cementitious compounds start to break down. This would also affect the geotechnical properties of the material.

Research has recently begun at TRL to address these issues, as part of a research project for the Highways Agency on the processing of contaminated land on highway sites. The work will consist of a series of laboratory leaching tests, followed by a pilot scale test fill and field trials. The material chosen for the investigation consists of a lightly contaminated silt/PFA mixture, to which varying quantities of a heavily contaminated sewage sludge have been added to produce a material more typical of a contaminated site. Both materials were obtained from the A13 Thames Avenue to Wennington site. The principal contaminants are metals, sulphate, sulphide and organic matter.

Initial tests have consisted of a series of trial mixes to produce an acceptable general fill material for use in the subsequent tests. The aim was to determine the mix with the highest proportion of sludge and the lowest proportion of lime which would give satisfactory geotechnical properties for a general fill; excessive lime content would render the process uneconomic. A mix with 5% sludge, 5% lime and 90% silt-PFA (bulk weight) was found to give satisfactory properties, these being defined as: an immediate CBR of greater than 3%; a 7-day soaked CBR of greater than 5%; a 28-day heave of less than 5mm average, with no one sample greater than 10mm.

The leaching tests are flow-through tests, carried out in specially designed triaxial cells to allow continual monitoring of the permeability of the samples. The pH and conductivity of the leachate are being monitored throughout the tests to assess the chemical stability of the materials. The leaching tests allow simulation of years of natural weathering in an embankment in a few weeks in the laboratory, and thus provide a guide to the long-term stability of the materials. Distilled water, dilute acid and diesel will be used as leachants to simulate various scenarios. Undrained triaxial tests will be carried out on the samples at the end of the leaching tests to assess any loss of strength compared to unleached samples.

Field trials are expected to begin in 1997, subject to finding available sites, and will utilise the results of the leaching and physical property tests. The results will enable an appraisal to be made of the long-term stability of lime-stabilised contaminated material in terms of both geotechnical properties and release of contaminants.

CONCLUSION

The research presented in this Paper has highlighted the variety of uses for quicklime. Recent publications on the use of lime for stabilised capping for highways can now assist designers in maximising the use of materials within the construction site. A draft report for using lime to improve bulk fill has been prepared and a specification is in draft form on the use of lime for track bed improvement. Trials for improving track bed and reducing ground vibration may be carried out next year, and lead to a more widespread use of the technique within railway maintenance strategies. The work on piles and columns, and processing contaminated land are less widely established but have considerable potential. The early results of this research are very encouraging.

The innovative use of lime has many benefits and as the UK becomes more accustomed to its appearance on construction sites, more applications will almost certainly be raised. These require investigation, because the processes are complex, and the specialist needs to be involved at an early stage to ensure the success of the project.

ACKNOWLEDGEMENTS

The work described in this study is part of the research programme of the Civil Engineering Resource Centre of the Transport Research Laboratory and was funded by the Road Engineering and Environmental Division of the Highways Agency and Railtrack. Many colleagues at TRL are thanked for their contributions to the Paper, in particular: Mr R Snowdon, Dr J M Reid, Dr J Temporal, Mr D Steele, Mr D Hiller, Mr F Yuille, Mr S Reynolds, Miss S McKnight, Dr D R Carder, Mr P Darley, Dr G West's (independent consultant) contribution to the lime piles and column section is much appreciated. Buxton Lime Industries are thanked for supplying the lime used in the testing programmes.

REFERENCES

British Standards Institution (1983), BS 1047, "Air-Cooled Blastfurnace Slag Aggregate for Use in Construction", BSI, London.

British Standards Institution (1989), BS 812, Part 124, "Testing Aggregates, Method for the Determination of Frost-heave", BSI, London.

British Standards Institution (1990a), BS 1377, "Methods of Test for Soils for Civil Engineering Purposes", BSI, London and as amended by British Standards Institution (1991), "Special announcement", BS 1377, BSI News, January 1991, BSI, London.

British Standards Institution (1990b), BS 1924 "Stabilised Materials for Civil Engineering Purposes", BSI, London.

Cripps, J C and Taylor, R K (1986), "Engineering Characteristics of British Over-consolidated Clays and Mudrocks: I Tertiary Deposits", Engineering Geology, 22, p.349-376.

Design Manual for Roads and Bridges, HMSO, London.
HA74, "Design and Construction of Lime Stabilised Capping" (DMRB 4.1.6).
HD25, "Foundations" (DMRB 7.2.2)
SH 7, "Specification for Road and Bridge Works: Soil Suitability for Earthworking - Use of the Moisture Condition Apparatus" (DMRB 4.1.4).

Glendinning, S and Rogers, C D F (1996), "Deep Stabilisation Using Lime", Lime Stabilisation, Edited by C D F Rogers, S Glendinning and N Dixon, Thomas Telford, London.

Harris, M R, Herbert, S M and Smith, M A (1995), "Remedial Treatment for Contaminated Land", Volume VII Ex-situ Remedial Methods for Soils, Sludges and Sediments, CIRIA Special Publication SP107, Construction Industry Research Information Association, London.

Highways Agency (1996), "Business Plan 1996/97", Highways Agency, London.

Johnson, P E (1985), "Maintenance and Repair of Highway Embankments: Studies of Seven Methods of Treatment", Department of Transport, TRRL Research Report 30, Transport Research Laboratory, Crowthorne.

Mackay, M and Emery, J (1993), "Practical Stabilisation of Contaminated Soil", Land Contamination and Reclamation, Vol 1, No 3, p. 149-155.

Manual of Contract Documents for Highway Works, Volume 1, "Specification for Highway Works", December 1991 with amendments, (MCHW 1), HMSO, London.

Mott MacDonald and Soil Mechanics Ltd (1994), "Study of the Efficiency of Site Investigation Practices", Department of Transport, TRL Project Report 60, Transport Research Laboratory, Crowthorne.

National Audit Office (1992), "Contracting for Roads", HMSO, London.

Nettleton, A, Robertson, I and Smith, J H (1996), "Treatment of Silt using Lime and PFA to form Embankment Fill for the New A13", Lime Stabilisation, Edited by C D F Rogers, S Glendinning and N Dixon, Thomas Telford, London.

Parsons, A W (1992), "Compaction of Soils and Granular Materials", a Review of Research performed at the Transport Research Laboratory, HMSO, London.

Perry, J (1989), "A Survey of Slope Condition on Motorway Earthworks in England and Wales", Department of Transport, *TRRL Research Report* 199, Transport Research Laboratory, Crowthorne.

Perry, J, Snowdon, R A and Wilson, P E (1996), "Site Investigation for Lime Stabilisation of Highway works", in *Advances in Site Investigation* (Ed. C Craig), Thomas Telford, London.

Rogers, C D F and Glendinning, S (1993), "Stabilisation of Embankment Clay Fills Using Lime Piles", Proceedings of the International Conference on Engineered Fills, Newcastle-upon-Tyne, Thomas Telford, London.

Sherwood, P T (1992), "Stabilised Capping Layers Using Either Lime, or Cement, or Lime and Cement", Department of Transport, *TRRL Contractor Report 151,* Transport Research Laboratory, Crowthorne.

Sherwood, P T (1993), "Soil Stabilisation With Cement and Lime", Transport Research Laboratory State of the Art Review, HMSO, London.

1	Plasticity Index and Grading - Within MCHW 1 requirements?	IF YES CONTINUE, IF NO REJECT
2	Establish Initial Consumption of Lime (ICL) - Is ICL established?	IF YES CONTINUE, IF NO REJECT

3	Total Sulphate Content / Total Sulphur Content / Organic Matter
	CBR Tests : (i) mellow before compacting
	                  (ii) 3 days curing
	                  (iii) followed by 4 days soaking
	                  (iv) then test
	Swelling, monitor to day 28
	Repeat tests at a range of moisture contents and lime additions from ICL in 0.5% stages
	Determine design lime addition value
	Is 7 day average CBR>15% (no individual specimen less than 8%)?
	Is average swelling ≤5 mm (no individual specimen >10 mm) and approaching an
	asymptotic value?
		IF YES, THEN MATERIAL IS
		SUITABLE FOR STABILISATION
		[subject to satisfactory water soluble
		sulphates in surrounding materials]
		IF NO REJECT

4	ADDITIONAL LABORATORY TESTS REQUIRED:
	        Test for frost susceptibility
	        Establish MCV/mc calibration for Class 7E material
	        Establish MCV/mc calibration for Class 9D material at design lime addition
	        Determine OMC for Class 9D material at design lime addition
	        Water soluble sulphate test (BS1377:Part 3 Amended in BSI News Jan 1991)	

Figure 1   Testing for soil suitability

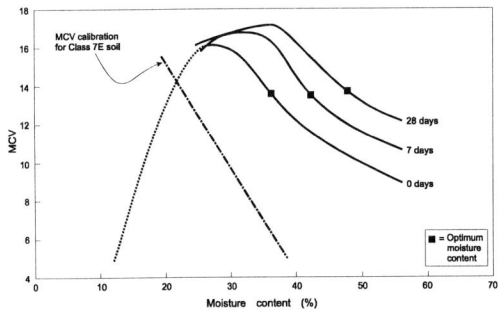

Figure 2   Effect of time on MCV and moisture content for a stabilised heavy clay

42

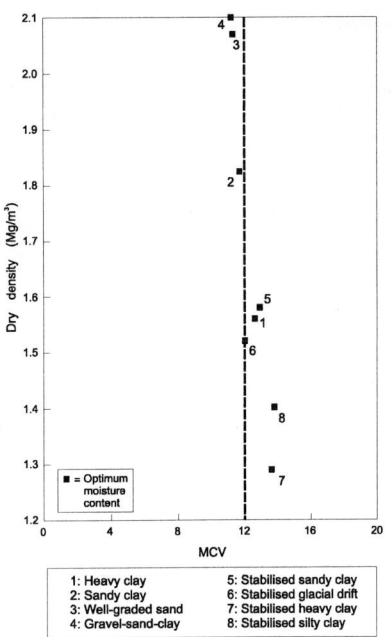

1: Heavy clay	5: Stabilised sandy clay
2: Sandy clay	6: Stabilised glacial drift
3: Well-graded sand	7: Stabilised heavy clay
4: Gravel-sand-clay	8: Stabilised silty clay

Figure 3   **MCV/Dry density (2.5kg rammer)/OMC relations**

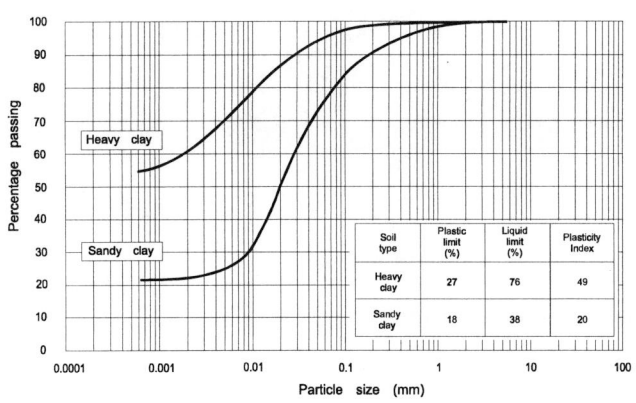

Figure 4   **Particle size distribution for cohesive soils**

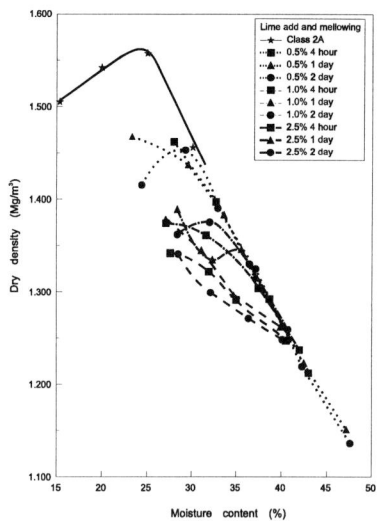

Figure 5   Dry density plots for treated and untreated heavy clay

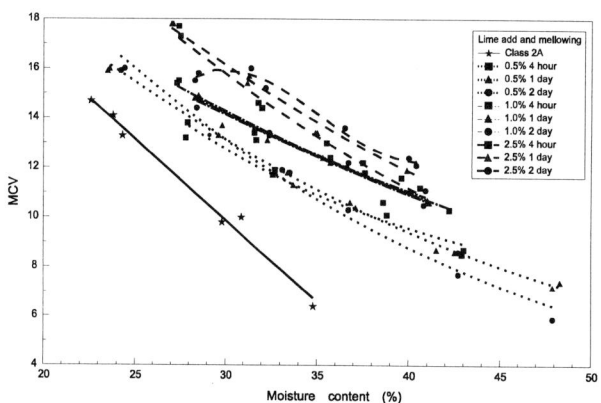

Figure 6   MCV calibrations for treated and untreated heavy clay

44

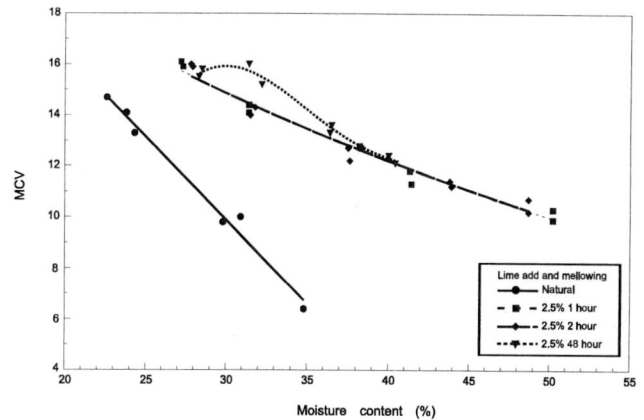

Figure 7   **MCV calibrations for lime treated and untreated London Clay**

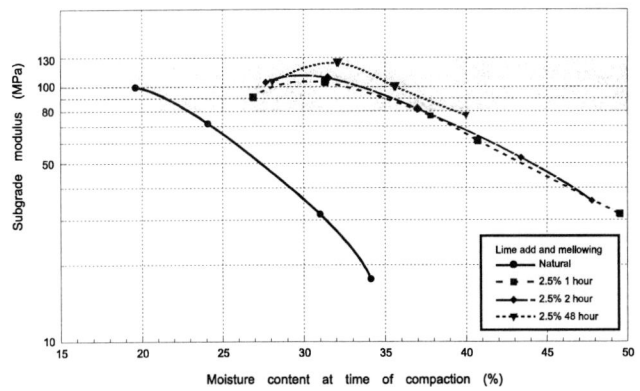

Figure 8   **Conversion of laboratory CBR data to subgrade modulus for lime treated and untreated London Clay**

SPECIFICATION AND PERFORMANCE OF LIME-CLAY MIXES

Introduction

Lime Treatment of Capping Layers under the Current DoT Specification for Highway Works

C C Holt BEng (Hons), PhD.
Formerly Research Student, School of Civil Engineering, Birmingham University.

R J Freer-Hewish PhD, BTech, CEng, MIEAust, MICE, MIHT.
Senior Lecturer, School of Civil Engineering, Birmingham University.

Long-Term Performance of Lime Stabilised Road Subgrade

S J Biczysko BSc, CEng, MICE, MIHT, MIAT, FGS.
Engineering Services Laboratory, Northamptonshire County Council.

The Structural Performance of Stabilised Soil in Road Foundations

B C J Chaddock BSc (Hons), PhD.
Transport Research Laboratory, Crowthorne.

Introduction

The traditional application of lime stabilisation is for the improvement of clay subgrades by mix-in-place techniques to provide foundations for road pavements. The improvements achieved by this method have often been found to be considerable, producing strengths and stiffnesses in excess of the more usual crushed rock aggregate foundation structures.

The primary role of a road foundation is to support a relatively small number of large stress applications caused by construction traffic during construction, and thereafter a very large number of low stress applications over a long period of time caused by vehicles riding on the completed pavement. The structural action required is one of load spreading to reduce the stresses transmitted to the natural subgrade to an acceptably low level, and sufficient inherent stiffness such that the complete foundation structure does not undergo transient deformations large enough to promote fatigue cracking of the overlying bound layers. In addition the materials must have sufficient strength to resist the build up of permanent deformation caused by the shear stresses acting during transient loading. Prediction of performance under loading is problematic due to the complex loading patterns applied, specifically in terms of varying magnitude and duration of stress, and stress direction (since the principal stress rotates as a vehicle passes overhead). A road foundation does, however, have another vital role, which is to provide support to the overlying layers during their compaction. Good compaction cannot be achieved on a soft underlying material.

Faced with this complex set of requirements, it is clear that the properties required both in the short term and the long term (i.e. after 20-40 years or more) are important. Added to this is the fact that the conditions within a road pavement structure are known to change with time, as drainage becomes effective and environmental conditions change (often worsen). It is for this reason that relatively poor equilibrium conditions are assumed for the natural subgrade in road pavement design.

The *modus operandi* of a lime stabilised subgrade is firstly compaction at or wet of optimum water content following mixing and mellowing. The material then progressively cements, becoming far stronger and relatively brittle. It is during this cementing process that the material is trafficked. It is common, therefore, for the material to crack, but this should not be considered to be detrimental as it is a natural action of the material. The material can thus be considered to be an aggregate containing large particles with a very high degree of interlock. In order to design and construct the stabilised layer adequately, therefore, it is important to know how to specify the construction process, how the material performs in both the short term and the long term, and what structural equivalence the performance has relative to current designs. An additional point that could be raised here is whether the enhanced performance noted from research on the lower layer(s) of the foundations render the material suitable to be used as the upper (sub-base) layer of the foundation. These issues are addressed by the papers in this chapter.

In the first paper, Holt and Freer-Hewish discuss a study of the effects of mellowing on the long-term performance of a lime stabilised clay. Mellowing, which is the period between mixing and compaction, was originally thought both to aid the pulverisation process, to the extent that it has been used as a pre-treatment followed by further lime addition, and to allow the modification reactions to take place fully. Although a period of 24-72 hours is commonly specified, depending on the clay type, there is little evidence that such a period is necessary and some strong evidence to suggest that it can result in a reduction in the maximum strength that can be achieved in the long term. It thus depends upon the reason for treating the clay as to whether a mellowing period should be used and, if so, of what duration to specify. If one is

48

considering long-term stabilisation, a short period would appear to be beneficial. If, as reported by Rogers and Glendinning in Chapter 3, the application is for treatment of soils for construction expediency (i.e. modification is the goal), then as long a period as is economic would appear to be better, particularly since low lime contents are generally used. Holt and Freer-Hewish examine the effects in laboratory tests on four British Clays.

The issue of long-term performance is addressed by Biczysko in the second paper. He reports a study of lime stabilised subgrades that were constructed approximately 15 years ago. A comprehensive programme of full pavement assessment *in situ* and of laboratory and field tests on material exposed in excavations is reported. The results are encouragingly good since, although some deterioration in the very much enhanced early performance (relative to untreated clay sections) occurred, the residual benefit was found to be clearly apparent.

The results of both of the above papers raise the question of whether a lime stabilised material could be used as sub-base. This is partially addressed in the third paper by Chaddock, who reports a study of lime and cement stabilised clay for use as sub-base. This study aimed to determine whether it was a feasible option by testing two basic structures with tapered sub-base layers on a relatively weak natural subgrade. The second question to be addressed, if the option is considered feasible, is how can such a layer be designed, i.e. what is its structural equivalency in relation to a standard Type 1 granular sub-base? These issues are examined with once again encouraging results.

The problem with any laboratory or controlled field study is in determining effects that realistic conditions have on long-term performance. This is addressed in part by the work of Biczysko. However, realism in laboratory testing for design purposes remains a challenge. The most effective tool is likely to be the repeated load triaxial test in a simplified form. Such a test remains a compromise since rotation of principal stresses is not an option, but it does give a means of determining accurately the resilient elastic stiffness and strength of the materials under an adequately large number of stress applications. Work of this nature has not apparently been carried out on stabilised British clays, although some work in the USA has been reported. Some opportunity exists to do a limited laboratory and field study as a result of a joint project currently underway between Loughborough University, Nottingham University and Scott, Wilson Kirkpatrick (Pavement Engineering) funded by the Highways Agency. It is hoped that such an extension to the work reported herein will provide the confidence needed to expand the use of lime stabilisation in pavement foundations.

Lime Treatment of Capping Layers in accordance with the Current Specification for Highway Works

C C HOLT AND R J FREER-HEWISH
School of Civil Engineering, University of Birmingham

INTRODUCTION

When naturally occurring soil is inadequate to form the subgrade of a pavement, several design options can be considered. These are:

I.   Accept the material as the subgrade with appropriate reinforcing layers

2.   Remove the material and replace it with a superior material

3.   Stabilise the existing in-Situ material to enhance its properties

Soil stabilisation techniques have become a popular and viable solution for improving the properties of inadequate soils in areas where good crushed aggregates are scarce. In addition, any use of the in-situ material reduces the wholesale demand for crushed aggregates for road construction and the associated haulage and waste disposal costs. Scarce resources are saved and the impact on the local environment is significantly reduced as a direct result of reduced truck movements. One method of soil stabilisation is the treatment of clay soils with small quantities of quicklime (CaO) or hydrated lime ($Ca(OH_2)$). Addition of lime to a suitably reactive soil will, within a few hours, produce a less plastic and more workable material. With time chemical reactions occur that strengthen and impWv~ the durability of the once marginal material.

THE SOIL-LIME REACTION PROCESS

When lime is added to a suitably reactive material two reaction phases occur. A reactive clay material is usually accepted as one with a clay content and plasticity index of 10% or higher. The first reaction occurs within several hours as a result of base exchange between ions of the soil and the lime. This causes a reduction in the diffused water layer and results in edge to face attraction of the clay particles (Hilt and Davidson, 1960). This reaction phase causes a flocculation of the soil particles and results in a modified material being produced which has a coarser texture, is more friable and is less plastic. Where the soil is excessively wet, drying due to the exothermic reaction results in a more workable material than in its original form. This part of the reaction process is more commonly known as the **modification** process. Immediate strength gains can also occur (Van Ganse,1973 and 1974), although there is evidence that these gains can be reversed.

When an alkaline environment with a pH value of around 12.4 is maintained after the immediate reactions have taken place, the pozzolanic reaction and **stabilisation** of the soil occurs. The high alkaline environment promotes the dissolution of silica and alumina from the clay particles which in turn reacts with the calcium ions from the lime to form calcium silicate hydrates (CSH) and calcium aluminate hydrates (CAH). These reaction products bond the clay particles together and are similar in composition to those of cement paste. Stabilisation of the material causes considerable increase in strength and durability, although the gain in strength is both temperature and time depend (Ruff and Ho, 1966).

SOIL-LIME STABILISATION FOR CAPPING LAYERS

On weak sub grades it is common practice in the UK to construct a capping layer between the subgrade and sub-base. This in effect reduces the thickness of the sub-base and provides a firm platform for compaction of the sub-base. Generally the capping layer is formed from granular material. However, influenced by the scarcity of aggregates, the 5th edition of the Department of Transport Specification for Road and Bridge Works (1976), gave the option of using a stabilised capping layer to improve the quality of the subgrade. This was the first time lime treatment was mentioned in the Department of Transport's specification, although lime treatment had been used successfully in the USA in the early 1950's and subsequently worldwide. It was not until the revised and renamed 6th edition of the Specification for Highway Works (1986), that a more detailed specification for cement and lime stabilised capping layers was included. From experience gained modifications were made and included in the current Specification for Highway works (1991/94) and the Design Manual for Roads and Bridges (1995).

CONSTRUCTION OPERATIONS

With the introduction of bigger and more powerful plant in the last 20 years the lime treatment process has improved dramatically since the early 1950's. Presently there are two construction methods adopted for lime treatment. These are the mix-in-place method and the static plant mix method, however the mix-in-place method is by far the most commonly used method in the UK and throughout the world.

All lime treatment for capping layers in the UK must be constructed in accordance with Clause 615 of the current Department of Transport's Specification for Highway Works (1991/94). A typical construction procedure adopted in the UK, in accordance with clause 615, is given below.

1. The formation must be adequately compacted and trimmed to the required profile and cross -section of the proposed treated layer.

2. Lime is then uniformly spread by mechanical means on the material *in situ*.

3. Mixing and pulverisation of the soil then takes place, after which the treated layer is lightly rolled and left to "mellow" for 24 to 72 hours. This mellowing period aids in the breakdown of large clay clods and is currently compulsory under paragraph 11 of clause 615.

4. After mellowing the layer is then remixed with sufficient water and compacted to ensure the air void content is 5% or less. This minimum air void content can normally be monitored by using an upper MCV value of between 12 and 14.

BACKGROUND TO "MELLOWING"

The construction process outlined above includes a compulsory mellowing period of between 24 and 72 hours before final compaction commences. The mellowing period is the time allowed to elapse between initial mixing of lime into the soil and final compaction of the treated material. However, no real scientific justification exists for the inclusion of mellowing in the current specification other than experiences gained in the USA in the late 1950's. Thomas *et al* (1965) reported that a mellowing period of 24 hours was required in order to meet pulverisation requirements on a project in Missouri in the late 1950's. Work carried out at the same time by Hoover (1965), in the state of Iowa, also suggested that a mellowing period of between 48 and 72 hours should be incorporated as a pre-treatment period before the addition of a second increment of lime immediately before compaction. McDowell (1959) reported that all treated soils should not be mellowed for more than 48 hours, although for some heavy clays a mellowing period of 96 hours may be required.

Two studies at that time, however, indicated the problems of long-term mellowing. One, undertaken by the Louisiana Department of Transport and reported by Taylor and Arman (1960), evaluated aspects of lime stabilised sub-base failures and concluded that treated layers that had received a mellowing period of more than 48 hours were much weaker compared to sub-bases that had received a mellowing period of less than 48 hours. This report led Mitchell and Hooper (1961) to investigate the effect of a 24 hour mellowing period on the properties of a lime treated expansive clay from California. Results suggested that densities decreased by up to 7.5%, cured strengths by 30% and swelling virtually doubled compared to specimens compacted after the immediate addition of lime. In the UK, at this time, Dumbleton (1962) considered the effect of mellowing on lime treated clays, although only as part of a much wider study, and concluded that the density of the lime treated material decreased with mellowing duration.

It was not until the late 1980's that further work was carried out. Sweeney et al (1988 and 1989) examined the effect of 1, 4 and 24 hour mellowing periods on a lime treated Regina clay from Saskatchewan Short- and long-term strengths were investigated and it was concluded that mellowing adversely affected the strength of the specimens.

RECENT RESEARCH AT BIRMINGHAM

Based on the fact that no real scientific evidence exists that support the inclusion of mellowing periods as part of the lime treatment construction process, a three year research programme was undertaken at the University of Birmingham. The research primarily investigated the lime treatment of four clay soils that are commonly encountered in subgrades in the UK and to assess whether lime treatment of these soils would be successful and how various mellowing periods affect the physical and engineering properties of the treated material. Secondly the effect of temperature and mixing moisture content were also examined to investigate the effect of these parameters on the mellowing period.

The four soils tested were Lower Lias, Oxford and London Clays and Mercia Mudstone (Keuper Marl). Some properties of these soils are characterised in Table 1. The lime used throughout the testing programme was quicklime with an available lime content of 96%. The quantity of lime added to each of these soils was equal to the initial Consumption of Lime value (ICL), BS 1924: part 2:1990. The ICL value was used so that the reactivity potential for pozzolanic reaction was similar for all soils. The Lower Lias, Oxford and London Clays required a 4% addition of quicklime, whereas only 3% was required for the Keuper Marl.

EXPERIMENTAL PROCEDURE

Two mixing regimes were used to simulate extreme site conditions. Firstly, quicklime was added to each soil in its dry state and was then mixed at the optimum moisture content (OMC) of the mixture and not the host material. The OMC used was that obtained after immediate addition of lime. This is important to note as results suggest that the OMC of the material alter with time. Secondly, quicklime was added to each soil in a very wet state. The moisture content present in the wet soil was equal to 1.2 times the Plastic Limit of the natural soil (Table 1). After lime mixing was complete the mixture was placed into a container and the surface was sealed by light tamping to emulate site conditions. The material was then placed into an environmental chamber set at either 5^0C, 10^0C or 20^0C at a relative humidity of 50% and mellowed for 0.5, 1, 2 or 3 days. The 5^0C temperature and 0.5 day mellowing period was included to determine the effects on the lime treatment process just outside the UK specifications. During mellowing the surface was left exposed to represent site procedures. After mellowing the modified compaction characteristics were determined and new batches of material were mellowed using the above procedure. After this mellowing procedure the lime treated specimens were brought back to the modified OMC and six strength specimens were prepared. These strength specimens were then cured for seven days at the mellowing temperature in a sealed condition. After curing three of the specimens were tested for strength and the remaining three were soaked for a further seven days before testing for soaked strength and volume change. Specimens were also prepared without mellowing so that comparisons with the mellowed samples could be made.

This paper only presents the results for a mellowing and curing temperature of 20^0C as this is the curing temperature specified in BS 1924 (1990), the current testing standard for cement and lime stabilised materials. This standard is generally used to ascertain whether a soil is suitable for treatment with lime.

Table 1   Properties of the materials tested

Properties of material	Lower Lias Clay	Keuper Marl	Oxford Clay	London Clay
Clay content (%)	64	37	57	54
Plastic Limit (%)	25	22	28	30
1.2xPlastic Limit	30	26.4	33.6	36
Plasticity Index	27	12	25	45
Maximum Dry Density (Mg/m^3)	1.595	1.827	1.534	1.583
Optimum Moisture Content (%)	21.5	16	24.2	24.3
UCS of untreated soil (unsoaked N/mm^2)	0.31	0.35	0.41	0.28
Total sulphate hate content (%)	1.12	0.11	0.88	0.23

RESULTS

Compaction Characteristics

Upon immediate addition of lime the optimum moisture content (OMC) increased and the)c maximum dry density (MDD) decreased for all four soils tested (Figures 1 and 2). These results have been reported by many researchers, such as Dumbleton (1962) and Mateos (1964). The reaction between the lime and the soil causes water to be hydrated from the soil. Flocculation and aggregation, as a result the soil-lime reaction, increases the air void content. These two reactions increase the soil's affinity for water and therefore an increase in OMC is recorded. Flocculation of the soil particles develops greater resistance to compaction and subsequently the MDD decreases for a given compactive effort.

Progressive increases in OMC accompanied by decreases in MDD were also observed for all soils with increased mellow duration, although the changes after a mellowing period of one day were slight, as shown in Figures 1 and 2. These changes to the compaction characteristics are due to the on-going reaction between the soil and the lime during mellowing. Greater aggregation occurs resulting in more water being lost and increased resistance to compaction. Experiments carried out by Diamond and Kinter (1965) and Verhasselt (1990) suggest that weak cementitious material is formed in the early stages of the lime reaction process. This would bind the clay particles together offering greater resistance to compaction.

Keuper Marl was the only soil that displayed a significant change between wet and dry mixing. However, the strength of all the materials does vary with mix conditions and is discussed later.

Unsoaked strength

Generally, for all dry mixed soils mellowed and cured at 20⁰C the maximum strength was achieved with no mellowing. Strength then generally decreased with mellowing duration as shown in Figure 3. This is a direct result of continued lime depletion during the mellowing period. As the mellowing period increases the reaction between the soil and lime continues. This continued reaction generates weak cementitious material that bonds the clay particles together producing a loose aggregated structure. Subsequent pulverisation and re-compaction breaks these relatively weak bonds, which cannot reform again. During the mellowing period available lime is consumed to satisfy initial reactions of the modification process. Subsequently less available lime is left to enter into the pozzolanic reaction stage that produces the cementitious material for strengthening and stabilising the soil. X-ray diffraction data (not included here) suggests that the cementitious gels produced with mellowing were less crystalline and structured than those with no mellowing.

Less lime was detected in specimens that had been mellowed. The reduction in strength can be as high as 20% after a one day mellowing period. However, reductions in strength after that were slight. All strengths measured were short term and it is unclear how mellowing affects the long-term strength of the treated clays tested, although from work carried out by Sweeney *et al* (1988 and 1989) it would appear that the strength reduction caused by mellowing may be magnified with prolonged curing.

There were, however, some exceptions. Both dry mixed Keuper Marl and Oxford clay gained maximum strength after a half day mellowing period before compaction. It is unclear why this occurred, but an optimum mellowing period does appear to exist for these materials.

The wet mixed soils demonstrated a somewhat different trend. Wet soils compacted after the ,immediate addition of lime produced the weakest strength as shown in Figure 4. In some cases the strength achieved was only slightly greater than the untreated clay compacted at its OMC. This was due to the initially high moisture content of the soil and the subsequent lower densities achieved. Considerable increases in strength, up to 100%, were recorded when the wet mixed lime treated samples had been mellowed for between a half and one day before specimen preparation and compaction. These short mellow periods allow enough water to hydrate from the soil-lime mix to allow compaction to take place at, or near the correct modified OMC.

For maximum strength gain it would appear that dry mixed specimens must be compacted immediately after the addition of lime, or within at least half a day of lime addition, depending upon the soil type. However, for wet mixed soils a mellowing period of between a half and one day is necessary to hydrate enough water to gain maximum strength.

Soaked strength

Three of the six cured specimens were soaked after curing for a further seven days. This testing regime may be considered harsh, but the possibility of "wetting up" of the mixes needed to be considered. The effect of lime treatment on the durability of the treated clay was therefore investigated.

With one exception, Keuper Marl (dry mixed), the strengths of all treated materials decreased A with soaking (Figure 3) Decreases were expected as a result of water permeating into the specimen. This causes a weakening of the cementitious material which leads to a subsequent reduction in strength. Decreases in strength after soaking also reflected the decreases in strength after mellowing (i.e. soaked strengths were generally greater with no, or relatively short mellowing periods). Dry and wet mixed Oxford Clay recorded the greatest reduction in strength, up to 93%, and increase in volume (refer below) after soaking. The strength of all treated clay was always greater than the strength of the soaked untreated soil as all untreated specimens collapsed after a relatively short soaking period (generally after approximately 3-6 hours).

Volume change upon soaking

The increase in volume after soaking was much greater for wet mixed specimens compared with dry mixed specimens. This is likely to be a result of the initial mixing process. It was noted that quicklime was not easily distributed throughout the wet mixed samples and large pockets of lime developed. After curing it was noted that cracks had developed where large pockets of lime occurred as a result of quicklime expansion. Subsequently this had forced the specimen apart and allowed water to penetrate the specimen relatively easily. The lump matrix of the wet mixed specimens was also considerably larger resulting in more air voids and a subsequent increase in permeability.

Oxford Clay experienced the greatest volume change. It has a relatively high sulphate content, which in itself would cause swelling. This was confirmed with X-ray diffraction analysis which indicated the presence of highly expansive ettringite and thaumasite at low temperatures. It is interesting to note that Lower Lias Clay with a larger sulphate content did not display exceptional swelling, however both Lower Lias and Oxford Clays demonstrated the greatest percentage reduction in strength after soaking. Keuper Marl, having the lowest sulphate content demonstrated very little volume change and percentage reduction in strength after soaking.

CONCLUSIONS

The research has addressed the effect of mellowing durations on lime treated materials that may be suitable for use in a capping layer.

The MDD decreased and the OMC of the mixtures increased with mellowing duration, but changes were only slight after a one day mellowing period. Moisture-density relationships constantly alter, and it is important that the correct moisture content, appropriate to the mellowing period, is available at the compaction stage. This ensures that there is enough water present for the soil-lime reaction to continue and enable the compacted treated layer to have an air void content of 5% or less.

Unsoaked strengths of all the materials investigated improved considerably after the addition of lime. In some cases increases in strength were considerable after a relatively short cure period.

Increased curing is likely to produce further strength gains. The potential for the lime treated material to be used as a sub-base or strong subgrade to eradicate the capping layer could possibly lead to further reductions in the overall construction cost of the highway.

Experimental work has revealed that prolonged mellowing can be deleterious to the strength gain process, although in certain cases it is desirable to mellow the soil-lime mix for up to one day. However, this is dependent upon the condition of the soil before treatment. It is therefore important that during the laboratory testing stage the expected mellowing period should be incorporated into the testing programme. This gives an indication of how moisture-density relationships are likely to alter with time. The effect of mellowing periods on the strength gain process can also be assessed for a particular soil. This will help to determine the optimum mellowing period for maximum strength gain, or indeed whether a mellowing period is required.

Dry mixed soils should be compacted immediately after mixing, with ideally not more than a half day delay for maximum strength and durability to be achieved. Wet mixed soils, however, require at least a half day mellowing period, but not more than one day, for maximum strength gains to be recorded.

REFERENCES

British Standards Institution (1990) "Methods of Test for Stabilised Soils", BS 1924, HMSO, London.

Department of Transport (1976) "Specification for Road and Bridge Works", Fifth Edition, HMSO, London.

Department of Transport (1986) "Specification for Highway Works", Sixth Edition, HMSO, London.

Department of Transport (1991) "Specification for Highway Works", Seventh Edition, HMSO, London.

Department of Transport (Amendment August 1994) "Specification for Highway Works", Seventh Edition, HMSO, London.

Design Manual for Roads and Bridges (1995) HA/74/95, Volume 4 Geotechnical and Drainage Works "Design and Construction of Lime Treated Capping Layers", HMSO, London.

Diamond, S and Kinter, E B (1965) "Mechanisms of Soil-Lime Stabilisation", Highway Research Record No 92, p. 83-102.

Dumbleton, M J (1962) "Investigations to Assess the Potentialities of Lime for Soil Stabilisation in the United Kingdom", Road Research Laboratories Technical Paper No 64, Crowthorne.

Hilt, G H and Davidson, D T (1960) "Lime Fixation in Clayey Soils", Highway Research Bull. No 262, p. 20-32.

Hoover, J M (1965) "Evaluation of Experimental Stabilised Soil Base Construction, Webster County, Iowa", Highways Research Record No 92, p. 21-42.

Mateos, M (1964) "Soil-Lime Research at Iowa State University", Proceedings of the ASCE 90 (SM 2), p. 127-153.

McDowell, C (1959) "Stabilisation of Soils with Lime, Lime-Flyash, and Other Reactive Materials", Highway Research Board, Bull. 231, p. 60-66.

Mitchell, J K and Hooper, D R (1961) "Influence of Time Between Mixing and Compaction on Properties of a Lime-Stabilised Expansive Clay", Highway Research Board No 304, p.14-31.

Ruff, C G and Clara Ho (1966) "Time-Temperature Strength-Reaction Product Relationships in Lime-Bentonite-Water Mixtures", Highway Research Record No 139, p. 42-60.

Sweeney, D A, Wong, D K H and Fredlund, D G (1988) "Effect of Lime on Highly Plastic Clay with Special Emphasis on Ageing", Transportation Research Record No 1190, p.13- 23.

Sweeney, D A, Fredlund, D G, Gan, J K and Widger, R A (1989) "Evaluation of Environmental Influences on a Lime-Modified Highly Plastic Soil", Paper Presented to RTAC Meeting, Calgary, Alberta, September 17-21, 1989.

Taylor, W H and Arman, A (1960) "Lime Stabilisation Using Preconditioned Soils", Highway Research Board, Bull No 262, p. 1-19.

Thomas, C E, Jones, W G and Davis, W C (1965) "Lime and Phosphoric Acid Stabilisation in Missouri", Highway Research Record No 92, p. 43-68.

Van Ganse, R F (1973) "Immediate Stabilisation of Wet Soils", Proceedings of the Eighth International Conference on Soil Mechanics and Foundation Engineering, Part 2, p.233-237.

Van Ganse, R F (1974) "Immediate Amelioration of Wet Cohesive Soils by Quicklime", Highway Research Record No 501, p. 42-53.

Verhasselt, A F (1990) "The Nature of the Immediate Reaction of Lime in Treating Soils for Road Construction", American Society for Testing Materials STP 1095, Philadelphia, 1990, p. 7-17.

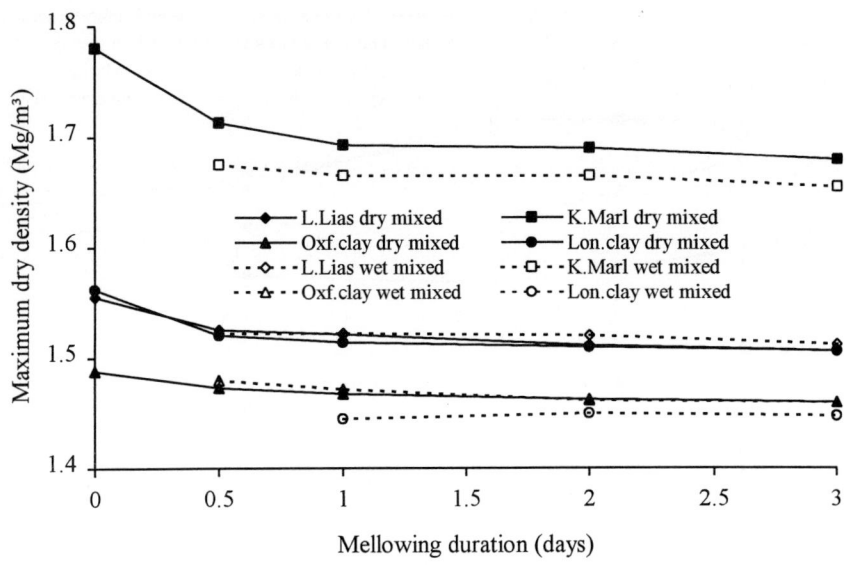

Figure 1   Changes to the MDD of all treated soils after mellowing at 20°C

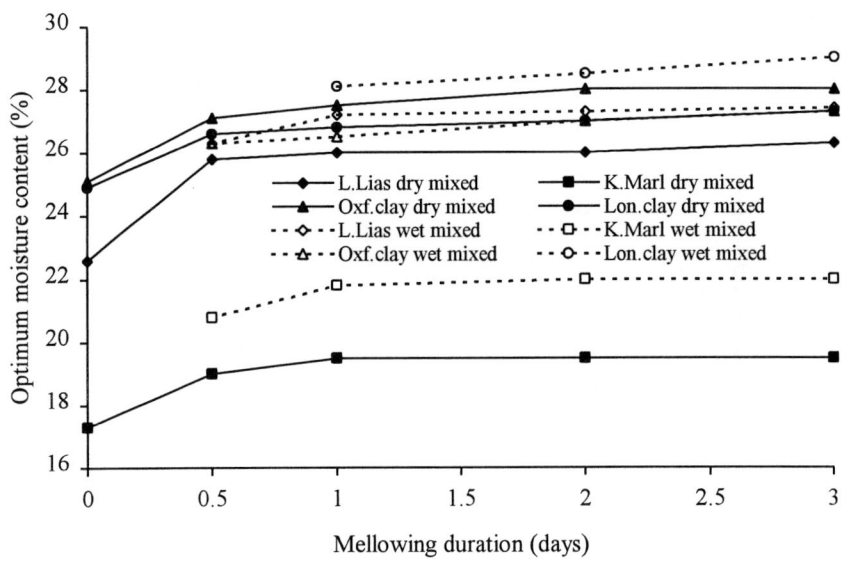

Figure 2   Changes to the OMC of all treated soils after mellowing at 20°C

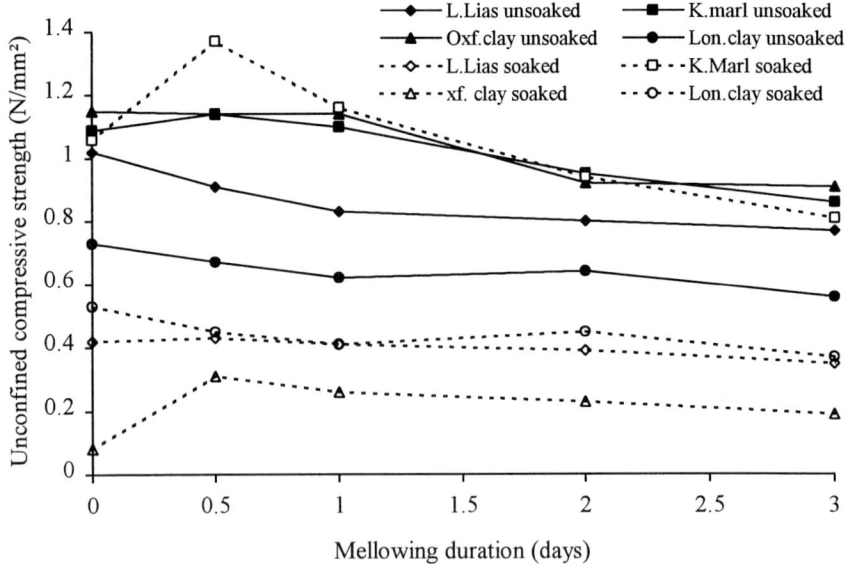

Figure 3 Changes to the unconfined compressive strength of all dry mixed soils after mellowing and curing at 20°C

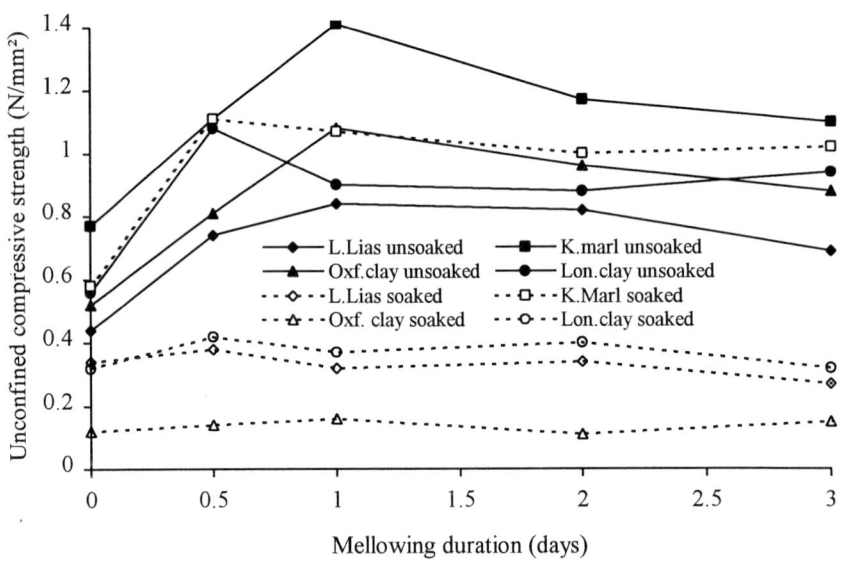

Figure 4 Changes to the unconfined compressive strength of all wet mixed soils after mellowing and curing at 20°C

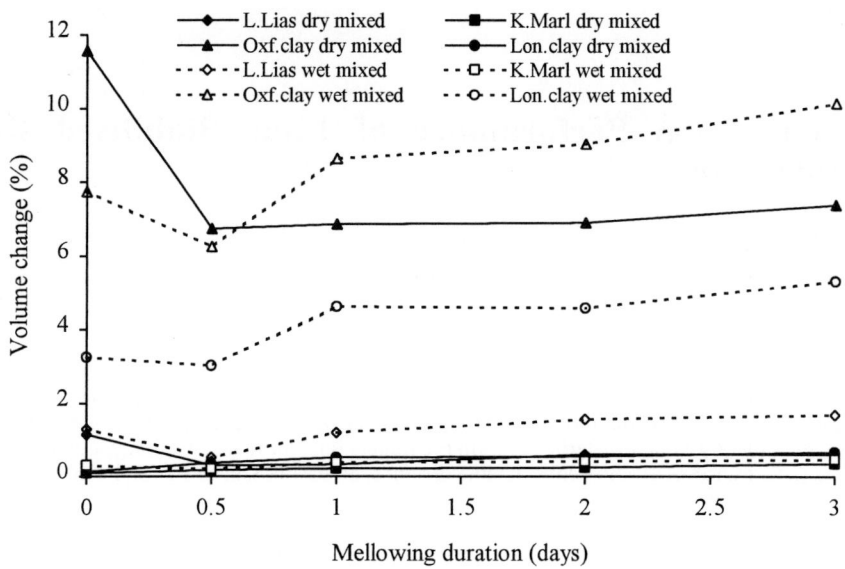

Figure 5 Volume changes to all lime treated soils after soaking

Long-Term Performance of Lime Stabilised Road Subgrade

S.J. BICZYSKO
Engineering Services Laboratory, Northamptonshire County Council

INTRODUCTION

Northamptonshire presents an area of varied geology derived from the Jurassic series of soils and rocks. None of the indigenous rock however provides characteristics suitable for the production of granular materials to form the foundations for roads. Consequently this product has to be transported some considerable distance, at least 60km, to any highway construction site in the area.

The prospect of improving the engineering performance of the natural soils is therefore attractive. This enables the soil to have an enhanced role in the structural characteristics of the road pavement. In-situ stabilisation was first considered as a means to achieve this objective in the late 1970s. Two projects were subsequently constructed, in 1980 and 1982, using lime stabilisation techniques. These schemes represent some of the earliest recorded use of lime stabilisation on the UK road network (Sherwood, 1992).

The road pavement structures in both locations have performed well over the intervening years. The sites afforded an opportunity for long-term monitoring and forensic study of the physical and engineering characteristics of the lime treated subgrade and road pavement over an elapsed service life of fifteen years.

STUDY SITES

In the early 1980s two road construction projects were undertaken using lime stabilised subgrade to improve foundation conditions. The sites of these projects were A45 dual carriageway at Ecton, Northampton (Location 1) and Brackmills Spine Road, Northampton (Location 2).

Location 1 consists of a section of dual carriageway road which was opened to traffic in 1981. The road at this point is on an embankment about three metres in height. The works contain a section, about 2,500m^2 in area, of lime treated subgrade. Little was known at the time about the characteristics and curing regime of lime in the UK environment although it had been used previously in overseas countries. The clay subgrade was treated using an MPH100 stabilisation machine to a depth of 300mm in a single layer. Lime was introduced, followed by light compaction and mellowing and thereafter repulverised and given final compaction. On one carriageway the soil was stabilised with quicklime whilst on the opposite carriageway hydrated lime was used. The lime application rate was approximately 3 per cent by mass in both cases and the works carried out at or close to optimum water content conditions. Adjoining sections of untreated clay soil provided control areas and the lateral extent of the lime treated area was extended beyond the carriageway itself into the verge and central reservation areas. The overlying road pavement was of composite cement bound material and bituminous surfacing. The completed road pavement surface was identified for the quicklime, hydrated lime and control sections. This enabled serviceability performance of the road to be examined by subsequent rolling wheel transient deflection surveys which are the usual method

Lime stabilisation. Thomas Telford, London, 1996

for in-service road pavement structural evaluation in UK. The strength gain of the lime treated soil was assessed at the time of construction by both triaxial and CBR methods. A rapid gain in strength from a CBR equivalent of 2 per cent to over 30 per cent was recorded. The road structure detail at Location 1 is illustrated in Figure 1(a).

Two years later, in 1982, a much larger project was undertaken at Location 2 at Brackmills, Northampton. The road is at, or close to, existing ground level. Quicklime at an application rate of 3 per cent for most of the works was used to stabilise the clay soil subgrade about 10,000m^2 in area. The details of this construction are described elsewhere (NCE, 1982). A composite cement bound material and bituminous surfacing formed the overlying structure of the road pavement. This location has been subject to periodic transient surface deflection surveys over the intervening years and includes a section of untreated subgrade as a control area. The stiffness of the lime treated subgrade has been measured *in situ* using static plate loading on several occasions through the service life of the road, and a forensic examination undertaken to establish the comparative condition and durability of the lime stabilised and control areas after a period of fourteen years. The road structure detail at Location 2 is illustrated in Figure 1(b).

SOIL CONDITIONS

The Jurassic soils and their derivatives which form the subgrade materials at both locations can be variable both in terms of physical characteristics and soil chemistry. These soils can contain high, and locally very high, concentrations of sulphates in the form of gypsum and selenite. The potential for subsequent expansive reaction in such soils in the presence of lime (Sherwood, 1962) ensured that attention was given at the site investigation stage to consider the soil chemistry. Sulphates were established at low levels before proceeding with the stabilisation works at both locations. However at Location 2 a limited area of acidic ground was revealed which required reworking and further addition of lime.

SERVICEABILITY

The serviceability of the whole road pavement structure including the stabilised subgrade and control sections was gauged by the transient deflection method. This process determines the surface deflection of the road under a 6350 kg rolling axle load. The deflection value is used to determine the residual life of the pavement, in standard 80 kN axles, or to derive the thickness of overlay strengthening necessary to extend structural life. At Location 1 the road surface was referenced with measurement points at 15m intervals and the transient deflection measured by a Deflection Beam (Kennedy *et al*, 1978) at each of the reference points. This survey was repeated at periodic intervals over ten years from construction. Correction of the recorded deflection measurements to a standard temperature condition of 20°C enabled comparison of the control section with both the stabilised quicklime and hydraulic lime areas. Typical deflection profiles for the quicklime stabilised carriageway are illustrated in Figure 2. The two year elapsed time profile illustrates the effect of the lime treatment when compared with the unstabilised area but at the five year time period the influence appears less apparent.

Environmental aspects, groundwater conditions, materials characteristics and traffic loading all contribute to a complex response condition for flexible composite roads over their service lives and therefore the data presented in Figure 2 may not be relevant to situations elsewhere. However, since the effects of lime treatment are the only variable at this site over the length of road pavement the transient deflection response forms a useful comparison with the adjoining untreated sections on a relative basis. A profile of average deflections for both the hydrated and quicklime sections is presented in Figure 3 together with data gleaned from the control sections over a thirteen year period.

The overall response of the road pavement appears somewhat better for the quicklime treatment in its early life period. This is perhaps due to the greater availability of lime, as CaO, in quicklime as compared with hydrated lime (Sherwood, 1993). However, in the longer term the performance of the quicklime and hydrated lime sections in terms of overall road pavement transient deflection appear to become convergent.

Flexible composite road pavements are of relatively stiff construction overall and the transient surface deflections are correspondingly low for a well constructed and maintained road pavement. This type of road is therefore sensitive to deflection change. However the deflection values throughout are of a low order for both the unstabilised area and the lime treated sections. On a comparative basis therefore the data is of value but all of the measured transient deflections are well below the level for intervention and all demonstrate long structural life expectancy.

After a service life of some ten years some exploratory excavations were made in the verge area of the dual carriageway. These excavations did not reveal the expected extension of the lime stabilised soils which were anticipated at subgrade level over the experimental area either at the margins of the road or in the central reservation area. This led to the prospect that in these areas the lime treated soil had been subject to carbonisation. The process of carbonisation is generally associated with concrete. However, the mechanism of change can also be applied to lime treated soils if atmospheric carbon dioxide could gain access to the treated area. Carbonisation studies of stabilised mixtures in a hot, arid environment (Bagonza *et al* , 1987) indicated that process reversion is possible. An examination of the condition of the lime stabilised subgrade directly beneath the dual carriageway was not feasible due to traffic management aspects. At location 2 however opportunities were more readily available for a forensic study of the subgrade conditions directly beneath the carriageway.

SUBGRADE CONDITIONS

The roads at Location 2 were constructed to enable infrastructure development of an industrial area to the south east of Northampton. Phased occupation of the area provided an opportunity to examine the road pavement over its service life of fourteen years to date. A section of unstabilised soil was retained as a control area and elsewhere the silty clay subgrade was stabilised with quicklime. Foundation construction was completed with a granular sub-base and the pavement structure comprised an overlying composite of cement bound material and bituminous surfacing. The lime treated road pavements have all performed well and retain significant structural life. Transient deflection techniques were used to provide the means of pavement structural evaluation and present a comparative method of examination of the lime treated subgrade with the control area. The transient deflection characteristics of relatively stiff composite pavement structures must be carefully evaluated especially in situations such as location 2 where the road construction was trafficked for some time before the full thickness of bituminous material was applied. However, the partial construction for the first few years of life enabled the stiffness characteristics of the lime treated soil to be evaluated directly. A ground investigation had been carried out for design purposes some time prior to construction and based upon measured data the road pavement thicknesses were derived from equilibrium CBR values between 2 per cent and 3 per cent. The ground investigation also confirmed that the site was suitable for lime stabilisation and established the virtual absence of sulphates. Prior to construction the subgrade soil response to various lime additions was evaluated. Immediately after completion of the stabilisation process specimens were extracted from the treated areas. With the exception of one limited area which had to be reprocessed by the addition of further lime to overcome an area of acidic ground the 3 per cent quicklime demonstrated an increase in CBR index to over 30 per cent in a matter of days. At the time when these works were undertaken the use of the Moisture Condition Value Test (Parsons and Boden, 1979) to assess earthworks treatment suitability was not in general use. The lime was introduced into the clay soil by MPH100 stabilisation machine in a single pass followed by a mellowing period, repulverisation and compaction to complete the process. The data gleaned from the ground investigation and construction monitoring testing is illustrated in Table 1.

Table 1 Summary test data for Brackmills Spine Road

Test Process	Water Content (%)	LL (%)	PL (%)	PI	CBR (%)	Shear Strength (KPa)	SO3	pH	Dry Density (Mg/m³)	Optimum Water Content (%)
Ground Investigation	23-27	63	28	35	2-3	75-100	Neg.	8.1	-	-
Pre-treatment	23-29	60	26	34	-	-	-	-	-	-
Plus 3% quicklime	29-32	-	-	-	20/29* 30/35#	-	-	-	1.76(max)	22
Construction validation	19-28	-	-	-	19/23* 28/35#	-	-	12.9	1.51-1.66	-

Note: * After two days # After seven days

The lime stabilised capping was introduced as equivalent to the design current at the time (Dept of Transport, 1978) and comprised the outline details illustrated in Figure 1(b).

MONITORING

A programme of rolling wheel transient deflection measurements at the road surface was undertaken for the first six years of service life in the stabilised areas and also in the natural soil control area. Summary values are illustrated in Table 2. During this period the final surface of the road had not been applied and the data therefore needs to be considered with caution. The deflection values however indicate a composite road pavement with good serviceability and adequate reserves of residual life.

Table 2 Average transient deflection (mm x 10⁻²) for Brackmills Spine Road

TREATMENT	ELAPSED TIME SINCE CONSTRUCTION (YEARS)		
	1	2	6
Quicklime (3 per cent)	15	11	17
Unstabilised control	12	-	18

The integrity and time related stiffness characteristics of the lime stabilised subgrade and control area was assessed using plate loading test techniques. Static plate loading consisted of a circular plate (300mm diameter) loaded through a calibrated load cell and load column acting against a kentledge reaction. The apparatus was set up in an exploratory pit which had been excavated through the road pavement construction to the upper surface of the subgrade. The plate was carefully bedded in contact with the subgrade and deformation measured by transducers at diametric points. The Modulus of Subgrade Reaction ('k') was determined (Day, 1976) as a means of characterising the stiffness of the subgrade at various time intervals. Plate loading test results are illustrated in Table 3 both for the lime stabilised subgrade and for the natural clay soil control area.

Table 3   Modulus of subgrade reaction (kN/m^2/mm) from plate loading tests and Brackmills Spine Road

PROCESS	PERIOD	ELAPSED TIME SINCE CONSTRUCTION (YEARS)					
		1		2		6	
LOADING		First	Second	First	Second	First	Second
Lime stabilised		148	–	138	220	+300	–
Lime stabilised		132	–	100	170	+300	+300
Unstabilised		49	–	–	–	67	108

The modulus values illustrate significantly enhanced stiffness of the lime stabilised soil compared with the control area of natural clay. Over time both the natural clay and the lime treated soil gained stiffness but the rate of increase for the lime stabilised material is significantly greater than for the untreated clay. The stiffness gain with time may be attributed to the material achieving equilibrium with the improved road drainage conditions and progressive pozzolanic strength gain within the lime stabilised soil over the six year measurement period. The k - value data may be interpreted through an approximate derived relationship with CBR index (Croney, 1991).

$$CBR = 0.01 \, k^{1.72}$$

Seasonal effects influence the k - value through water content changes but after one year the CBR index of the natural clay soil was of the order of 8 per cent increasing to in excess of 15 per cent after a period of six years. Equivalent data for the lime stabilised soil illustrates a rise in CBR index from 50 per cent after one year, to 60 per cent after two years and in excess of 100 per cent after six years. The CBR however can only be considered as an index of performance. The parameter has some value in that there is considerable experience in use and familiarity with the derived data but the test itself is a complex combination of elastic stiffness and deformation. Likewise although there is no specified k -value for acceptability a threshold of 65 kN/m^2/mm has been recognised as providing a threshold for a suitable granular foundation platform for construction.

FORENSIC INVESTIGATION

At Location 1 a number of exploratory pits were formed in the verge area adjacent to the lime stabilised carriageways after an elapsed service life of fourteen years. At the time of construction it was intended that the lime treatment be extended beyond the carriageway into the verge and central reservation earthworks. There was however no obvious evidence of lime treatment in these areas in the exploratory pits. Undisturbed and disturbed specimens were taken from the exploratory pits in both verges on a depth profile within the expected zone of lime stabilisation and undrained shear strength and pH value established. Details are presented in Figure 4.

The data appears to suggest a relationship between pH and undrained shear strength. However, there is some degree of uncertainty about this prognosis since it is not known definitively that the area was formerly lime treated although the higher pH values, which are in excess of the background level would suggest that it may have been. The soil water movement in the verge areas differ from that beneath the carriageway where the subgrade is effectively waterproofed by the road pavement structure and the road drainage. Since it was not possible at location 1 to continue the investigation beneath the carriageway to establish the long-term properties of the lime stabilised subgrade a forensic study was undertaken at location 2. Keyhole evaluation techniques were used to obviate the need for exploratory pits which disrupt the road pavement structure and can be troublesome to restore effectively. This examination was carried out after a service life of fourteen years by comparing the characteristics of two lime stabilised areas with the control natural soil. The process of keyhole evaluation comprised the formation of core holes of either 100mm or 150mm in diameter through the bituminous and cement bound layers of the road structure. The coring was terminated upon reaching the upper horizon of the road foundation and a portable Dynamic Cone Penetrometer (Figure 5) inserted into the core hole and advanced through the road foundation into the underlying subgrade. After the penetrometer test was completed a window sampling device was advanced through the same strata to reveal the details of foundation and soil profile and to provide samples for subsequent test. The DCP provides a measure of resistance to advancement of a 20mm diameter 60° cone by an 8kg sliding mass falling through 575mm. The resistance to advancement, or Cone Penetration Index (CPI) expressed as blows per mm, provides a relative measure to determine the properties and performance of road pavements (Kleyn and Savage, 1982) and this can be correlated with CBR index parameters through relationships such as that described by Kleyn and Van Heerden (1983)

$$\text{Log CBR} = 2.48 - 1.057 \text{ Log (CPI)}$$

There is generally a good agreement over most of the CPI range but for low values of CBR index, especially in fine grained soils, the data needs careful analysis. The gradient of the DCP advancement illustrates layer thicknesses and changes with depth which were confirmed by measurement from the soil profile in the window sampler. An illustrative DCP profile from Location 2 is presented in Figure 6.

Specimens of the soil profile from the window sampler were subject to organoleptic checks and phenolphthalein indicator tests (BSI, 1990) to establish the presence and extent of the lime treated soil. A series of laboratory tests including plasticity index, water content, sulphate, pH and strength indicator was carried out on specimens from both the control and stabilised areas. These data are presented in summary format in Table 4. The original ground investigation and construction information obtained fourteen years previously and illustrated in Table 1 may be compared with the information in Table 4. The lime stabilised subgrade continues to afford engineering characteristics appropriate for a stable foundation. The pH value of the lime stabilised soil appears to have reduced from a construction value in excess of 12 to a mean value of 10.9 over the intervening fourteen years. It is not clear if this is significant in engineering terms as a phase of slow carbonisation which will decay to the background level pH value of 8.2 given sufficient time. Unless deterioration of the foundation occurs, and there is no evidence to suggest this, then the road should continue to

Table 4   Summary test data: lime stabilised subgrade soil - service life 14 years

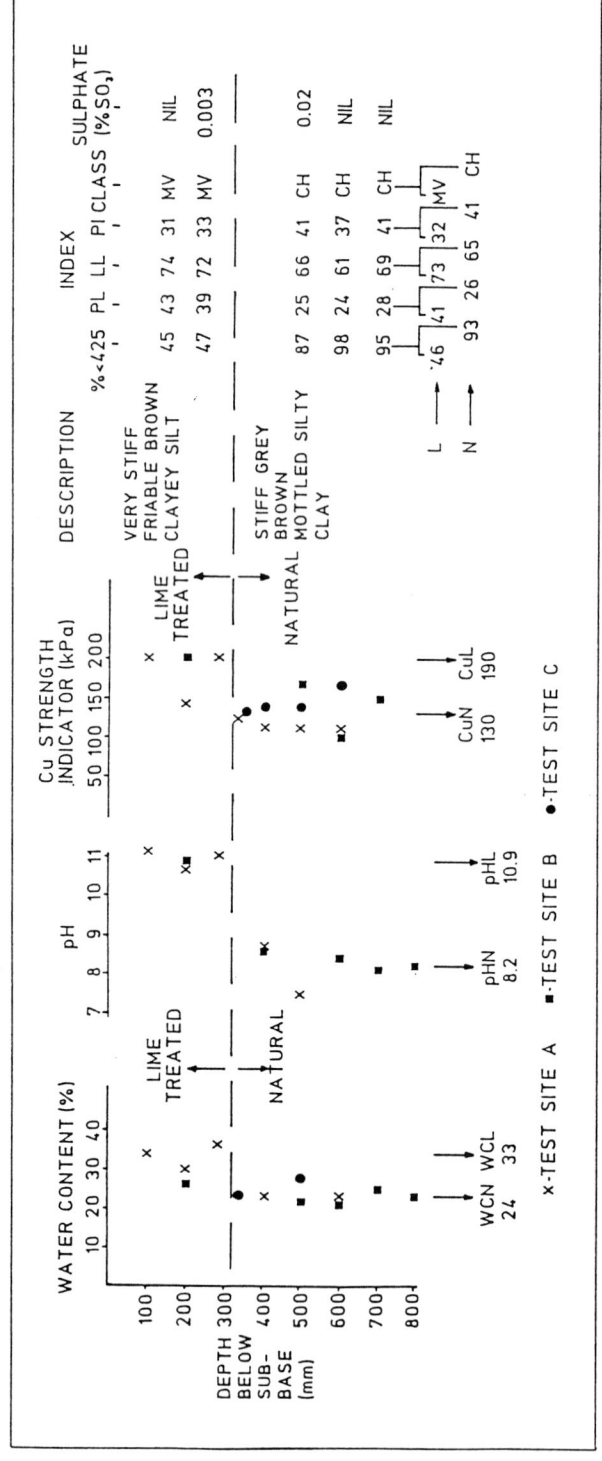

give good serviceability.

The role of the subgrade, in overall pavement performance, providing a sufficient threshold stiffness is maintained, does not have the same significance in long-term serviceability as compared to the immediate support provision at the time of construction. A displacement of soil plasticity from clayey to silty behaviour in the lime treated soil appears to have occurred with time. Plasticity characteristics are known to be altered by the introduction of lime (Dumbleton, 1962; Cobbe, 1988). Much of this plasticity change information relates to the immediate or short-term condition for the lime and soil mixture. The observed reduction in percentage finer than 425 micron over the fourteen year time period is attributed to pozzolanic effects in the lime treated soil which influences both Liquid and Plastic Limits and results in a displacement on the plasticity chart as illustrated in Figure 7.

STIFFNESS PARAMETERS

The foundation for a road has to provide appropriate performance both at the time of construction and throughout its service life. For the foundation the greatest stresses occur at the construction stage. It is also important to mobilise sufficient stiffness characteristics throughout the service life allowing for change in the environmental conditions with time. The evolved designs for thickness and materials are based largely upon empirical index methods based upon previous performance. The CBR index is not a direct measure of stiffness modulus, shear strength or deformation, but a complex interaction of all with variable response dependent upon material type. Various correlations exist relating elastic stiffness parameters to index tests such as CBR (Powell *et al*, 1984) but care must be taken to ensure that these are only used within the range of their applicability. The lime treated soils examined in this study were of a very friable nature and present significant difficulty in extracting an 'undisturbed' specimen of aged material for subsequent testing in the laboratory to obtain a measure of elastic stiffness. Plate loading *in situ* offered a possible alternative but requires an exploratory pit to carry out the test thereby obviating the advantage of the keyhole evaluation. The Dynamic Cone Penetrometer data as illustrated in Figure 6 can be interpreted to yield elastic modulus parameters (Chua, 1988) for the granular sub-base material, the lime stabilised subgrade and the underlying natural soil. These derived parameters are presented in Table 5.

Table 5   Stiffness parameters derived from DCP data

Layer	Elastic Modulus (MPa)
Granular Sub-base	480
Lime soil (upper)	241
Lime soil (lower)	158
Clay soil	55

The derivation of elastic modulus data provides useful engineering parameters which enable analysis of the overall multi-layer road pavement structure by analytical methods. This is helpful in the context of lime stabilised materials to illustrate performance relative to granular materials in terms of load spreading efficiency.

CONCLUSIONS

Since the construction of these two lime stabilised projects sixteen years ago much has been learned about the behaviour of such processes in the UK environment. This long-term investigation has confirmed the durability and effective service of the lime stabilised subgrade at both sites. Possible effects of carbonation could benefit from closer scrutiny in the longer term. The potential for lime stabilisation requires careful evaluation of the ground conditions (Perry *et al* , 1996). Providing an exhaustive investigation is undertaken and the construction operation thoroughly executed and validated then the characteristics of the lime stabilised material should provide good in-service performance as illustrated by this long-term study over the previous sixteen years.

ACKNOWLEDGEMENTS

The author is indebted to M.J. Kendrick OBE, CEng, Director of Planning and Transportation, Northamptonshire County Council for his permission to publish this paper. Thanks are also due to fellow Engineers who knowingly, or unwittingly, may have contributed towards the content, views and observations described, and to colleges at ESL for assistance and support in the testing and preparation of the text. The project described at Location 2 was undertaken for Northampton Borough Council as engineering agent to the former Northampton Development Corporation. The accuracy of statements remain the responsibility of the author and should not be construed as policy on behalf of his employer.

REFERENCES

Bagonza, S, Peete, J M, Freer-Hewish, R and Newell, D (1987), "Construction of Stabilised Soil Cement and Soil Lime Mixtures", Proc. Seminar H, PTRC Transport and Planning Annual Meeting, Univ. of Bath.

British Standards Institution (1990), "BS1924 : Part 2, Methods of Test for Cement Stabilised and Lime Stabilised Materials", BSI, London.

Chua, K M (1988), "Determination of CBR and Elastic Modulus of Soils Using a Portable Pavement Dynamic Cone Penetrometer", Penetration Testing 1988, ISOPT-1, De Ruiter(ed), ISBN 90-6191-801-4, Rotterdam.

Cobbe, M J (1988), "Lime Modification of Kaolinite-Illite Clays", Civil Engineering Technology, February 1988, p.9-15.

Croney, D and Croney, P (1991), "The Design and Performance of Road Pavements", ISBN 0-07-707408-4, McGraw Hill, p.168-171.

Day, J B A (1976), "Proof Testing of Unbound Layers", Proc. Int. Symp. on Unbound Aggregates in Roads (UNBAR1), Univ. of Nottingham.

Department of Transport (1978), "Technical Memorandum H6/78 : Road Pavement Design", Table 7S.

Dumbleton, M J (1962), "Investigations to Assess the Potentialities of Lime for Soil Stabilisation in UK", Road Research Technical Paper No.64, HMSO, London.

Kennedy, C K, Fevre, P and Clarke C S (1978), "Pavement Deflection: Equipment for Measurement in the UK", Report LR834, Transport Research Laboratory, Crowthorne.

Kleyn, E G and Savage, P F (1982), "The Application of the Pavement DCP to Determine the Bearing Properties and Performance of Road Pavements", Proc. Int. Symp. on Bearing Capacity of Roads and Airfields, Trondheim, Vol 1, p.238-242.

Kleyn, E G and Van Heerden (1983), "Using DCP Soundings to Optimise Pavement Rehabilitation", Proc. Annual Trans. Convention, Johannesburg, Report LS/83 Materials Branch, Transvaal Roads Dept., Pretoria, S. Africa.

New Civil Engineer (1982), "Lime Stabilises Subgrade", 23.09.82, Thomas Telford, London.

Parsons, A W and Boden, J B (1979), "The Moisture Condition Test and its Potential Application in Earthworks", Supplementary Report 522, Transport Research Laboratory, Crowthorne.

Perry, J, Snowdon, R A and Wilson, P E (1996), "Site Investigation for Lime Stabilisation of Highway Works", Advances in Site Investigation Practice, ISBN 0-72-772513-0, Thomas Telford, p.85-96.

Powell, W D, Potter, J F, Mayhew, H C and Nunn, M E (1984), "The Structural Design of Bituminous Roads", Laboratory Report 1132, Transport Research Laboratory, Crowthorne.

Sherwood, P T (1962), "The Effects of Sulphates on Cement and Lime Stabilised Soils", Roads and Road Construction, Vol 40, No 470, p.34-40.

Sherwood, P T (1992), "Stabilised Capping Layers using either Lime, or Cement or Lime and Cement", Contractor Report 151, Transport Research Laboratory, Crowthorne, p.45.

Sherwood, P T (1993), "Soil Stabilisation with Cement and Lime", ISBN 0-11-551171-7, HMSO, London.

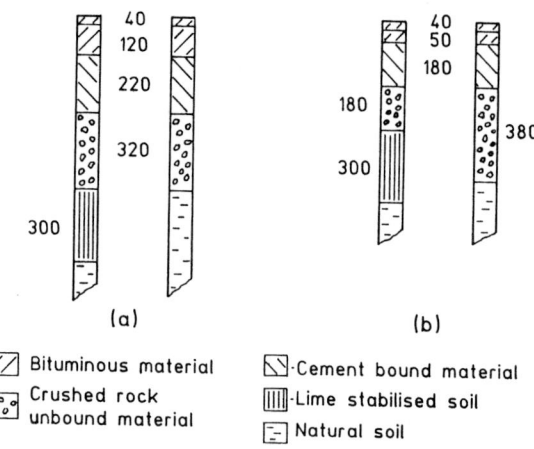

Figure 1   Outline structural details (a) of A45 at Ecton and (b) Brackmills
Spine Road.

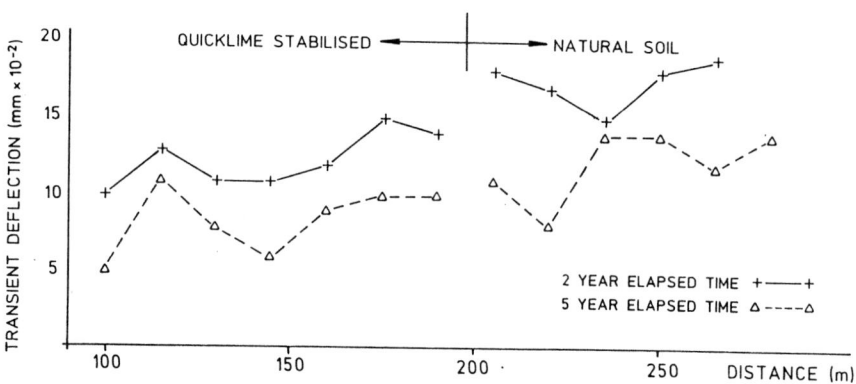

Figure 2   Transient deflection profiles of A45 at Ecton

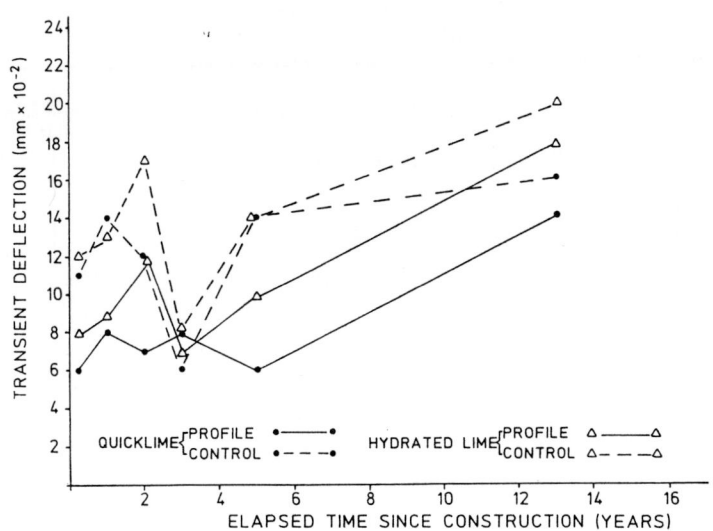

Figure 3   Surface Deflection - Time relationship of A45 at Ecton

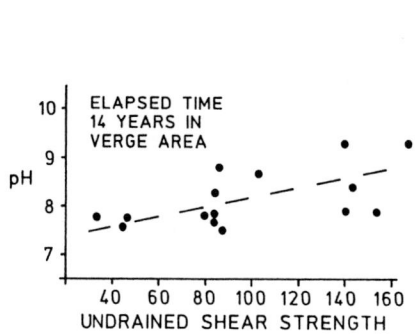

Figure 4   Conditions in road verge at A45, Ecton

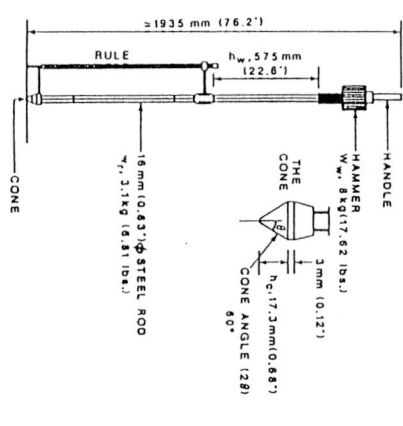

Figure 5   Dynamic Cone Penetrometer

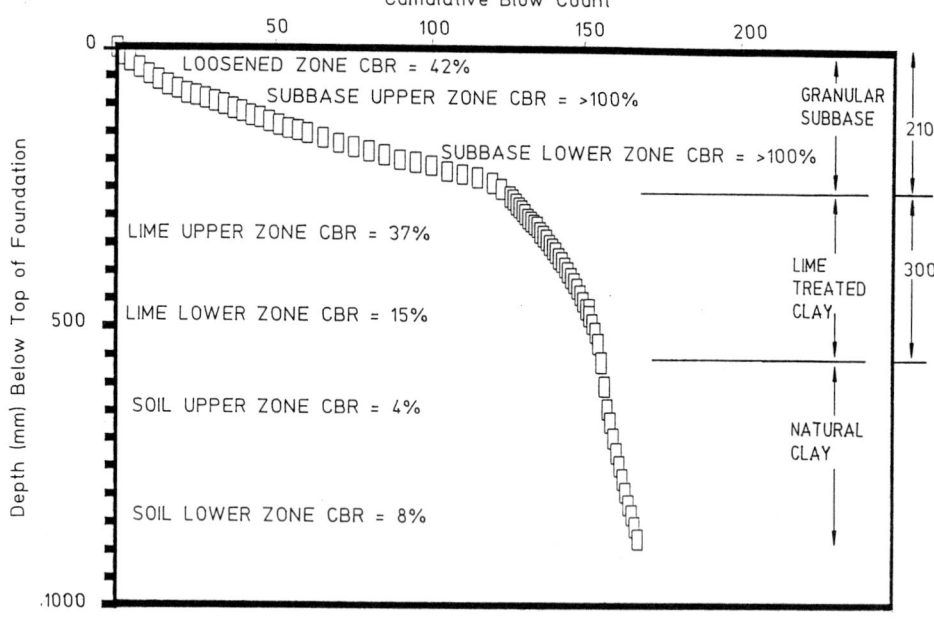

Figure 6 DCP depth profile in foundation of Brackmills Spine Road

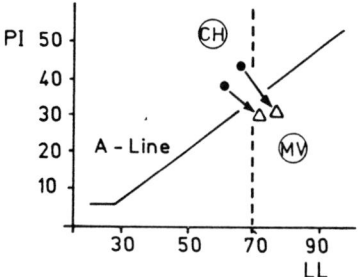

Figure 7 Plasticity change in subgrade of Brackmills Spine Road after 14 years

The Structural Performance of Stabilised Soil in Road Foundations

B.C.J. CHADDOCK
Transport Research Laboratory, Crowthorne

INTRODUCTION

In principle, foundation design consists of choosing the thickness of sub-base appropriate for the soil subgrade strength and foundation performance required. All other factors being equal, it should be permissible for sub-bases to be laid thinner, the more superior their structural properties. However, specified thicknesses of Department of Transport unbound granular, Type 1 sub-base and stabilised sub-base given in HD 25 (DMRB 7.2.2) are the same regardless of sub-base strength. Also, the Department of Transport limits stabilised sub-bases to cement bound materials of strength no lower than that of Type CBM1 material. Consequently, the purpose of the research described here is to widen the range of materials accepted as stabilised sub-base. In addition, it is required to determine the thicknesses of various stabilised sub-bases that are structurally equivalent to Type 1 sub-base as this should allow the stronger sub-bases to be laid thinner than the traditional and widely used Type 1 sub-base. This work should therefore lead to both economic and environmental benefits.

The research reported is concerned with cohesive soil that is initially treated with lime and then cement. This work, however, is part of a larger project that also includes cement stabilised granular materials. The complete project forms one element of the LINK Transport Infrastructure and Operations Programme and is supported by British Cement Association, Buxton Lime Industries, County Surveyors' Society and the Department of Transport. The whole project is showing clearly the uses and benefits of a Portland cement binder that needs to be considered alongside those of lime that are emphasised in this paper.

The paper considers the role and behaviour of sub-base layers in road foundations. It then outlines the mix designs of the stabilised soil sub-bases, and describes the construction of full-scale trials consisting of these sub-bases and a control Type 1 sub-base laid directly on a soil subgrade. An account is given of the structural testing of these foundations that was primarily carried out to assess their performance as working platforms for pavement construction. The stabilised soil sub-bases were shown to perform well in comparison with Type 1 sub-base. Designs for stabilised soil sub-bases are recommended.

ROLE AND BEHAVIOUR OF SUB-BASE LAYERS

The structural properties of a sub-base material that control its performance in the road are stiffness, strength and resistance to deformation. Stiffness is important because, when traffic loads the sub-base, this layer must have a sufficiently high stiffness to prevent the stresses transmitted through the sub-base exceeding those values that can be sustained by the underlying weaker materials. This behaviour is especially important during road construction when the stresses imposed directly on the sub-base are higher than those applied through a completed pavement. The sub-base must also contribute to the stiffness of the completed foundation, which must be sufficiently high to allow good compaction of the roadbase and to limit flexing and cracking of the bound pavement layers by traffic on the finished road. In addition, the sub-base must have sufficient strength and deformation resistance to perform these functions without suffering excessive deterioration. Stabilised sub-bases and unbound granular sub-bases deteriorate in different manners.

During road construction, foundations comprising unbound granular sub-bases on clay subgrades deteriorate by rutting when trafficked. Repeated loading by construction traffic can cause shear deformation within the subgrade and sub-base resulting in the sub-base thinning beneath the loaded wheels and heaving outside the wheel paths. The nature of the deformation was revealed by trenches dug through foundations by Chaddock (1988). Large ruts can form in inferior foundations and the subgrade can break through the heaved sub-base adjacent to the wheel paths in extreme cases.

Foundations comprising stabilised sub-bases crack by shrinkage and thermal contraction; especially during their first night after construction. These cracks are more closely spaced and therefore narrower, the weaker the sub-base. This type of cracking should not be considered as deterioration as it is a natural behaviour of this material. However, allowance for these cracks should be made in design. Construction traffic can cause deterioration of stabilised sub-bases in the vicinity of naturally occurring cracks. The extent of deterioration is dependent on the nature and amount of the construction traffic, the subgrade support and the load transfer across the cracks; a factor that relies on the degree of aggregate interlock. A combination of coarse aggregates and high temperatures closing narrow cracks results in good aggregate interlock and load transfer. Conversely, a combination of fine grained aggregates and low temperatures opening wide cracks results in poor aggregate interlock and load transfer. Under traffic, pumping can occur at cracks with the development of voids beneath the sub-base. Stabilised sub-bases can then cantilever and crack. Construction traffic can crack weak, thin stabilised sub-bases that are inadequately supported. Repeated trafficking can then lead to deformation in the wheel paths. In addition, at cracks, frost can damage the sub-base and fretting can occur under traffic.

The aim of foundation design is to choose the thickness of a selected sub-base that is appropriate for the soil strength and envisaged traffic and which results in the foundation performing satisfactorily without excessive deterioration.

STABILISATION PROCESS OF LIME FOLLOWED BY CEMENT

The effect of lime and cement used individually and in combination on soils is described by Sherwood (1993). It was concluded that the addition of lime and then cement to a soil can combine beneficial properties derived from both materials.

Lime prepares the soil for cement. For a wet soil, this is achieved by the combined effect of drying the soil by absorption and evaporation of water and changing the plasticity characteristics of the soil by the lime modification process. For drier soil requiring added water, the lime treatment normally reduces the plasticity of the soil. The overall effect of these changes is to produce a more workable material by breaking down large clods of clay into a looser, finer and more friable material.

Dumbleton (1962) showed that higher strengths at seven days were obtained by cement treatment compared with lime treatment when a sandy clay and a silty clay were separately mixed with the same quantities of lime and cement. For a heavy clay, similar strengths were obtained. Consequently, cement treatment of soil prepared with lime is more likely to produce a material with sufficient early life strength to be used as a sub-base than treatment of soil by lime alone. In the longer term, lime introduced directly into the soil, or arising indirectly from the cement, may increase the strength of treated soil containing pozzolanas by the lime stabilisation process.

OUTLINE OF FULL-SCALE TRIALS

Full-scale trials were carried out to widen the range of permissible sub-base materials and to

76

determine the thicknesses of various stabilised sub-bases that are structurally equivalent to Type 1 sub-base. Trial sections were constructed in the TRL Pavement Test Facility (PTF). They were each of length 10m and width 2.4m and were laid on an imported soil subgrade of California Bearing Ratio (CBR) value between 2.5 and 3.0 per cent. A control section of unbound granular Type 1 sub-base and two sections comprising soil treated with lime and then cement were built. In order to obtain information on a wide range of sub-base materials, low and high strength versions for the stabilised soil were constructed. The minimum strength for the weaker material (CLS-L) was a CBR value of 30 per cent. For the stronger material (CLS-H), a cube crushing strength was required that was as high as possible without using uneconomic quantities of lime and cement. Each sub-base was constructed as a wedge with its thickness varying along the trial section to give a range of performance. The stronger sub-base (CLS-H) was laid thinner than the weaker sub-base (CLS-L). The structural properties of the stabilised sub-bases were established in the laboratory by determining how their strength varied with age. The structural performance *in situ* of each foundation was assessed by simulating the traffic the foundation would carry during road construction and monitoring its deterioration by cracking, deformation and change in stiffness.

MATERIALS

Subgrade

The subgrade was constructed from London Clay that was imported into a pit in the PTF. The soil had a liquid limit of 72 per cent, a plastic limit of 26 per cent and a plasticity index of 46 per cent. According to Casagrande's extended soil classification system, the soil is described as Type CH, a clay of very high plasticity. The subgrade for each trial section was shaped, with increasing depth below pit edge along the section, to allow the construction of a wedge of sub-base material.

Measurements of the subgrade strength were made *in situ* with the MEXE Cone Penetrometer that was fitted with a 20mm diameter cone (Black, 1979). Tests were conducted on the exposed subgrade prior to construction of the sub-bases, down core holes and at the bottom of trenches in the sub-bases formed after trafficking the foundations. The average strength of the subgrade in each trial section is recorded in Table 1. Measurements of the resistance to penetration of the cone over a depth of 750mm are given together with CBR values. CBR values were estimated from cone penetrometer tests using a previously determined calibration given by CBR = 0.036*CI. In this equation, CI is the average cone index over a depth of 75mm determined from tests *in situ* and CBR is the laboratory CBR value measured on "undisturbed" specimens.

Table 1 Subgrade strength

Subgrade for:	Resistance to penetration of cone (CI scale) over 750mm		CBR (%) at top of subgrade	
	Before sub-base construction	After trafficking	Before sub-base construction	After trafficking
Type 1	81	86	3.0	2.9
CLS-H	86	102	2.5	3.9
CLS-L	83	103	2.5	4.2

Before the sub-bases were constructed the top 75mm of the trial sections had similar strengths with CBR values between 2.5 per cent and 3.0 per cent. At the end of the experimental programme, the CBR of the top of subgrade beneath the control Type 1 sub-base was unchanged but the strengths of the formations to the stabilised sub-bases CLS-H and CLS-L had increased to CBR values of 3.9 per cent and 4.2 per cent respectively. This strength increase is also evident from the results of the cone penetrometer tests when they are averaged over a greater depth of 750mm.

Sub-base

In the full-scale trials, sub-bases comprising an unbound granular crushed rock and a stabilised soil were laid on the clay subgrade. The Type 1 sub-base material used for the control section was a crushed granite from Mountsorrel, Leicestershire. A soil was imported from Telford, Shropshire for stabilisation. In its natural state, the soil had a Liquid Limit of 34 per cent, a Plastic Limit of 15 per cent and a plasticity index of 19 per cent. According to Casagrande's extended classification system, the soil is described as type CL, a clayey silt of low plasticity.

Selection of lime content. For the stabilised soil sub-bases, the primary purpose of adding quicklime to the soil was to prepare it for cement. This practical requirement was to be achieved by adding sufficient lime to the natural soil to produce a material of moisture content drier than its optimum moisture content (OMC). Adding cement to the lime treated soil in this condition would then require wetting the material up to its OMC; a process which was easier to perform than drying the material back to the specified compaction moisture content. For convenience, the moisture contents of the soil at various stages in its treatment were assessed by the Moisture Condition Value (MCV) test. According to HA 74 (DMRB 4.1.6), the MCV at OMC of various natural and lime stabilised soils varies between 11 and 14. An MCV of over 14 is therefore likely to be on the dry side of OMC.

To estimate the quantity of lime to produce a lime treated soil of MCV over 14 from stockpiled soil, MCV tests were performed on soil prepared to various moisture contents and treated with different lime contents of 1, 2 and 3 per cent. The MCV tests were carried out 24 hours after addition of the lime to simulate the effect of the mellowing period *in situ* . It was found that the addition of 2 per cent of lime was sufficient to modify a natural soil of MCV 5 to a lime treated soil of MCV 14. The stockpiled soil sampled for these tests had an MCV of 6.1. The addition of 2 per cent of lime could therefore be expected to modify the soil to a material with a MCV of just over the target value of 14, or a moisture content slightly dry of OMC. Consequently, a lime content of 2 per cent was provisionally chosen.

A secondary purpose of adding quicklime to the soil was to provide suitable conditions for the lime stabilisation process. Consequently, the minimum quantity of lime required to maintain reaction between the lime and any reactive components in the soil was determined. This value is known as the Initial Consumption of Lime (ICL) and is the minimum amount of lime needed to produce a solution of pH 12.4 at a temperature of 25°C. The ICL for this soil was determined according to the British Standards Institution (1990) and was 2 per cent, which gave support to the value previously selected.

After treating the soil with 2 per cent of lime and leaving to mellow for 24 hours, the soil was found to have a Liquid Limit of 47 per cent, a Plastic Limit of 23 per cent and a plasticity index of 24 per cent. The Plastic Limit, Liquid Limit and plasticity index of the soil all increased with addition of lime. This behaviour was considered by Buxton Lime Industries (1990) as being the consequence of silt dominating the plasticity characteristics of the modified soil.

Compaction characteristics of soil treated with lime and cement. To maximise strength and durability, the lime and cement treated soil should be compacted to its peak dry density. Compaction characteristics of this material were therefore determined in laboratory tests carried out immediately after the addition of cement to simulate conditions *in situ* . For soil modified with 2 per cent of lime and stabilised by 6 per cent cement, the OMC was found to be 15 per cent at a peak dry density of 1860 kg/m^3. When the cement content was changed to 10 per cent, the OMC was 14 per cent and the peak dry density 1880 kg/m^3. The MCV of both of these mixes was 12 and was within the range of 11 to 14 given in HA 74 (DMRB 4.1.6) for natural and lime treated soil. Consequently, it was decided to compact the lime and cement treated soil for the mix design tests at an MCV of 11.5 to produce a material that was slightly wet of its OMC.

Selection of cement content. Laboratory compressive cube and CBR tests were carried out to determine the quantities of cement that must be added to the lime treated soil to produce the required seven day strengths. For these tests, three quantities of cement, 2, 6 and 10 per cent, were mixed with the lime treated soil after 24 hours mellowing together with sufficient water to produce materials of MCV 11.5. Cubes of size 150mm were prepared for the two higher cement content mixes and tested in compression at seven days. Specimens for CBR tests were prepared for all three cement content mixes and also tested at seven days in unsoaked and soaked conditions. Unsoaked specimens were sealed for seven days whereas soaked specimens were cured in this manner for three days prior to soaking the specimens for four days. CBR values were required for the weaker materials for comparison with the minimum laboratory CBR value of 30 per cent specified for Type 1 sub-base in HD 25 (DMRB 7.2.2.). Cube compressive strengths of the stronger materials were required for comparison with the strengths of traditional, cement bound granular materials (MCHW 1, Series 1000).

The average compressive strengths after seven days were 1.05 and 1.40 MPa for soil treated with 2 per cent lime and bound by 6 and 10 per cent of cement respectively. Based on these results, a material treated by 2 per cent lime and bound with 8 per cent cement was expected to have a compressive strength of 1.25 MPa at seven days. This strength was considered to be satisfactory for the higher strength version of the stabilised soil (CLS-H) and did not use excessive quantities of binders.

The CBR value of the stabilised soil varied with cement content (C) according to the following relationships:

Unsoaked CBR = 14.3*C + 29.6
Soaked CBR   = 14.6*C + 33.5

The predicted CBR value of the soil treated with two per cent of lime and bound with 8 per cent cement was 144 per cent for unsoaked specimens and 150 per cent for soaked specimens. These results confirm the high strength of this material (CLS-H) and demonstrate its durability.

As the minimum laboratory CBR value of Type 1 sub-base is 30 per cent, the target CBR of the lower strength version of the stabilised soil was selected to be slightly higher than this value at 35 per cent to 40 per cent to reduce the risk of producing material for the full-scale trial that did not comply with minimum design requirements. It was predicted that soil treated with 2 per cent of lime would have a CBR of 51 per cent in the unsoaked condition and 55 per cent in the soaked condition when bound with 1.5 per cent of cement. Previously, MSTS Ltd had carried out tests on this soil when it was treated with 1.5 per cent lime and bound with either 4 per cent or 5 per cent cement. Their work suggested that a decrease in the MCV of the soil of 1.0 would reduce its soaked CBR value by about 10 per cent. Consequently, it was decided to produce the lower strength stabilised soil at a MCV of 10.5 so that this material would have an unsoaked CBR value of about 40 per cent.

The mix designs were:

For CLS-H: 2.0 per cent of Quicklime and 24 hours mellowing.
             8.0 per cent of Ordinary Portland Cement.

Moisture Condition Value of 11.5 at compaction.
For CLS-L;   2.0 per cent of Quicklime and 24 hours mellowing.
1.5 per cent of Ordinary Portland Cement.
Moisture Condition Value of 10.5 at compaction.

THICKNESS DESIGN OF EXPERIMENTAL SECTIONS

A foundation should be designed so that it acts as a satisfactory working platform for road construction without sustaining excessive damage by construction traffic. For foundations of unbound granular, Type 1 sub-base laid on soil subgrades, Powell *et al* (1984) considered a deformation of 40mm in the sub-base surface to be the maximum that can be tolerated if the sub-base is to be reshaped and recompacted efficiently and serious rutting is to be avoided in the subgrade. They proposed thicknesses of sub-base for a range of soil strengths and amounts of construction traffic that restricted sub-base deformation to below the 40mm criterion. For a subgrade complying with the target CBR of 2.5 per cent, a sub-base thickness of 350mm was recommended to carry 1000 standard axles of construction traffic. Consequently, it was planned to lay the Type 1 sub-base in a wedge whose thickness tapered from 425mm to 325mm to allow for variations of the subgrade strength from its target strength, as well as differences between the mechanical properties of the sub-base used in the PTF and those of the sub-bases used in the trials reported by Powell et al (1984). The thickness design of the Type 1 sub-base is given in Table 2 together with the thicknesses chosen for the stabilised soil materials.

Table 2   Design thicknesses of sub-bases

Material	Thickness (mm), minimum to maximum
Type 1	325 to 425
CLS-H	225 to 325
CLS-L	350 to 450

The thickness design for the lower strength version of the stabilised soil (CLS-L) was chosen to be very similar to that of the Type 1 sub-base as the target CBR value of this stabilised material was not much greater than the minimum CBR value of 30 per cent required of Type 1 sub-base. The stronger version of the stabilised soil (CLS-H) was chosen to be laid 100mm thinner than the thicknesses specified for the Type 1 sub-base.

CONSTRUCTION OF SUB-BASES

Powerbetter Developments Ltd were contracted to produce the lime and cement treated soil materials. The equipment used was a Bomag MPH 120. It was connected to a tanker so that water could be delivered within the hood encasing the rotating head to ensure good mixing of water, binder and soil. A mat of soil of thickness about 0.6m was laid on hard standing outside the PTF and the top 0.35m of the soil was processed.

The MCV of the natural soil was measured to be 6.8 and therefore the soil was treated with 2 per cent of lime. The lime was delivered in bags and spread by hand within a frame whose area was calculated to produce the required lime concentration within the processed depth.

The lime was then rotovated into the soil and the degree of pulverisation of the lime treated soil was determined according to the British Standards Institution (1990) as 65 per cent, which demonstrated that the material was adequately fragmented. The lime treated soil was then compacted by a roller and covered with sheets to protect it from weather during the mellowing period.

After 24 hours mellowing, the average MCV of the lime treated soil was 11.8 and the material was therefore considered suitable for the addition of cement. Two adjacent strips on the mat were assigned for production of the two stabilised soil sub-bases. The stronger sub-base, CLS-H, was processed and laid in the PTF prior to the production and compaction of the weaker sub-base, CLS-L. Bagged cement was spread in frames over areas calculated to achieve the target cement contents of 8 per cent for CLS-H and 1.5 per cent for CLS-L. The cement was thoroughly mixed into the lime treated soil by the stabilising equipment. Water was added to the materials on a trial and error basis. The measured MCVs of CLS-H and CLS-L were 11.2 and 10.6 respectively and were in good agreement with the target values. Both versions of the lime and cement treated soil were shown to be adequately fragmented by their degree of pulverisation that was determined to be at least 65 per cent. The processed material was removed by excavator from the top of the mat with care being exercised to avoid picking up unprocessed soil.

All three sub-bases were transported into the PTF by dump truck, tipped, spread and compacted in two layers with a Bomag BPR 80/60D reversing plate. The number of passes of this equipment was the same as that specified for the same thickness of Type 1 sub-base (MCHW 1, Series 800). When the layer varied in thickness along the trial, the compaction required for the thicker part of the layer was applied over the complete layer. For each stabilised soil sub-base, the surface of the bottom layer was loosened after compaction before laying the upper layer to help bond together these layers.

The bulk density of each sub-base was determined using a nuclear density gauge (NDG) operated in the transmission mode. For Type 1 sub-base, the moisture content was also measured with the NDG and its dry density calculated. In Table 3, the dry density of the Type 1 sub-base measured *in situ* is compared with its peak dry density that was measured in the compactibility test for graded aggregates specified by British Standards Institution (1980) and previously reported by Chaddock *et al* (1995). The Type 1 sub-base was adequately compacted as its in-situ dry density was 94 per cent of the peak dry density.

Table 3   Compaction of unbound granular sub-base

Sub-base	Depth of in-situ probe (mm)	Dry density (kg/m^3)		Degree of compaction: In-situ / Peak (%)
		In-situ	Peak	
Type 1	300	2094	2220	94

It was not necessary to compute dry densities for the stabilised materials as compaction to refusal tests were carried out in CBR moulds at the same time as the in-situ compaction of the sub-bases. As recorded in Table 4, the average in-situ bulk densities were at least 95 per cent of the mean bulk densities of specimens compacted in CBR moulds. By extension of the specification for the cement bound granular materials (MCHW 1, Series 1000), it can be seen that the stabilised soil sub-bases were adequately compacted.

Table 4   Compaction of stabilised soil sub-bases

Sub-base	Depth of in-situ probe (mm)	Bulk density (kg/m^3) for:		Bulk density ratio: In-situ / CBR (%)
		In-situ tests	CBR specimens	
CLS-H	250	2048	2138	96
CLS-L	300	2003	2119	95

The compacted thicknesses of the sub-bases along the trial sections are shown in Figure 1. The minimum and maximum thicknesses of the sub-bases differ by less than 6 per cent from their corresponding target values given in Table 2.

STRUCTURAL PROPERTIES OF SUB-BASE

Specimens of the stabilised sub-bases were made from material used to construct the trial foundations and their strength measured at different ages. Compressive cube strength tests were carried out on specimens of the higher strength version (CLS-H) of the stabilised soil. CBR tests were performed on both strength versions of the stabilised soil. The average strengths of the stabilised materials at 7, 28 and 91 days are given in Table 5.

Table 5   Strength of stabilised soil materials

Sub-base	Compressive cube strength (MPa) at age (days):		
	7	28	91
CLS-H	1.7	3.0	3.8
	Unsoaked CBR (%) at age (days):		
	7	28	91
CLS-H	138	191	205
CLS-L	36	45	58

The strengths of all the stabilised materials were in good agreement with the seven day target strengths. Hence, materials with markedly different strengths were produced for full-scale assessment. The variations of the compressive cube strengths and CBR values of the sub-bases with time are shown in Figures 2 and 3 respectively and increased by at least 50 per cent between 7 and 91 days.

Durability tests specified in MCHW 1 (Series 1000) were conducted on the higher strength version of the stabilised soil (CLS-H). The average compressive strength at 14 days of cube specimens immersed in water was 2.1 MPa compared to 3.0 MPa for the control specimens; a reduction of 32 per cent that exceeded the 20 per cent allowable reduction specified for cement bound granular material. However, this result conflicts with the results of CBR durability tests carried out as described in HA 74 (DMRB 4.1.6) and shown in Figure 3. These results show that the soaked CBR values were greater than the unsoaked CBR values at seven days for both versions of the stabilised soil. At 28 days, the strengths of these sub-bases were only reduced by about 7 per cent by soaking for 25 days. No reason for this conflicting durability assessment has been found.

STRUCTURAL BEHAVIOUR OF FOUNDATIONS

The structural performance of each foundation was assessed by measuring its stiffness with the Falling Weight Deflectometer (FWD) and monitoring its deterioration under traffic.

Stiffness

The FWD, that has been described by Sorensen and Hayven (1992), was used to measure stiffness *in situ*. In these tests, it was used as a dynamic plate bearing test in a similar manner to that described by Chaddock and Brown (1995). They tested foundations comprising unbound granular materials laid on soil subgrades, and showed that their stiffnesses were dependent on plate size and applied stress. Consequently, all stiffnesses reported are specific to the diameter of the loading plate adopted and have been corrected to values applicable to standard stresses. Tests were conducted on the subgrade at formation level and at the top of the sub-base and gave values of stiffness known as formation stiffness and foundation stiffness respectively.

An average value for the formation stiffness of each trial section is recorded in Table 6. These values were derived from tests with the FWD fitted with a 450mm diameter plate and were corrected to an applied stress of 60kPa. The magnitude of these stiffnesses are low, as would be expected of this weak clay. Also, the subgrade on which the two stabilised sub-bases were placed was less stiff at the time of construction than the subgrade on which the Type 1 sub-base was laid. These formation stiffnesses act as a reference against which to judge the contribution of each sub-base to foundation stiffness.

Table 6   Formation stiffness

Sub-base:	Type 1	CLS-H	CLS-L
Formation stiffness (MPa)	14.5	10.9	9.3

The stiffness of each foundation was measured at stresses that more closely approached those imposed by wheel loads. In these tests, a loading plate of diameter 300mm was adopted because, under this smaller plate, the derived foundation stiffness was more influenced by the upper part of the foundation and, hence, the sub-base. Measurements of foundation stiffness were corrected to values applicable to a surface stress of 400kPa. Tests were performed on the stabilised sub-bases just prior to the commencement of trafficking at seven days. Foundation stiffness is shown plotted along the trial sections in Figure 4 and as a function of sub-base thickness in Figure 5. The foundation built with the CLS-H sub-base has the highest stiffness and is followed in order of decreasing stiffness by the foundations constructed with CLS-L sub-base and the Type 1 sub-base. The superior performance of the stabilised soil sub-bases occurred despite these materials being laid to either the same thickness as Type 1 sub-base, or thinner than this unbound granular material. This behaviour occurred because the stabilised sub-bases were stiffer than unbound granular sub-base and the structural properties of the soil at the top of subgrade beneath the stabilised sub-bases had improved after placing these materials. For each sub-base, the foundation stiffness increased with sub-base thickness. The stiffnesses of the foundations incorporating stabilised sub-bases were the maximum possible at this age for these materials and foundation designs because the sub-bases were uncracked.

Performance under traffic

The performance of the foundations under traffic was assessed by repeatedly applying loads to the foundations in the PTF with a half axle wheel assembly fitted with twin tyres. Wear of the foundations was assessed by the magnitude of deformations in the wheel paths, the extent of cracking of the stabilised foundations and the change in foundation stiffness. Trafficking commenced on the seventh day for the stabilised sub-bases and was in two phases. The number and magnitude of the loads applied to each foundation are schematically illustrated in Figure 6.

During the initial phase of testing, all three foundations were trafficked by wheel loads of the same magnitude. Also, equivalent patterns of trafficking, in terms of the number of wheel loads applied at specific days after construction, were adopted for the foundations built with stabilised sub-bases. The first wheel load magnitude approximated a target load of 40kN. When applied to a full axle, this load is equivalent to a standard axle of 8.16tonnes. The magnitude of the wheel loads was then increased to include firstly 51.5kN and then 56.5kN. On a full axle, these loads respectively equate to the maximum axle loads of 10.5tonnes and 11.5tonnes that are currently permissible and will be allowed in the year 1999 respectively.

Of the stabilised sub-bases, only the lower strength version of the stabilised soil, CLS-L, was cracked during the initial phase of trafficking. The development of a multiple crack pattern should be considered normal for this material whose CBR value was designed to be similar to that of Type 1 sub-base. For the higher strength version of the stabilised soil, CLS-H, it was necessary to induce cracks in the sub-base during the final phase of trafficking so that the behaviour of a cracked sub-base could be studied. As illustrated in Figure 6, this stabilised soil was cracked by abnormal loads of 95kN on a half axle or an equivalent axle load of about 19 tonnes. After cracking this sub-base, the foundation was trafficked by at least 250 passes of a 5.75 tonne wheel load. No deterioration was observed at the cracks. The extent of the cracking after trafficking is illustrated schematically in Figure 7. The applied loads were converted into standard 8.16 tonne axles using a fourth power wear law, that is commonly adopted in pavement design, and their cumulative sum after completion of trafficking is given in units of a thousand standard axles in Figure 6.

After completion of trafficking, foundation stiffness was measured by the FWD and its variation along the trial section is shown in Figure 8. The change in foundation stiffness during the trials is revealed by comparison of these results with measurements of foundation stiffness made at seven days and shown in Figure 4. Only two positions could be tested on the Type 1 sub-base due to the large ruts and trial pits. These results indicated that foundation stiffness at the thick end of the sub-base layer did not change significantly during the study. However, the stiffnesses of the foundations constructed with stabilised soil sub-bases decreased during the trial as a result of the cracks induced in these sub-bases by trafficking and temperature changes. At the ends of the trial sections where the sub-bases were thickest, the foundation built with CLS-H had the highest stiffness after trafficking and was followed in order of decreasing stiffness by the foundations constructed with CLS-L and Type 1 sub-base. As recorded in Figure 7, only the stabilised soil, CLS-L, was visibly cracked at this end of the trial sections. At the other end of the trial sections, where the sub-bases were thinnest, both stabilised foundations were cracked. Although, the stiffness of the foundation built with CLS-H sub-base had markedly reduced, its stiffness was still greater than the stiffness of the foundation constructed with CLS-L sub-base.

In these trials, the robustness of the foundation built with the higher strength version of the stabilised soil, CLS-H, was demonstrated by the need to use abnormal loads to crack the sub-base. However, during road construction, cracks will occur as a result of shrinkage combined with reduced overnight temperatures, especially during the first night after construction. Consequently, the design of a stabilised foundation should be based on its behaviour in a cracked state.

EQUIVALENT THICKNESSES OF SUB-BASES

In order to reduce the design thicknesses of strong stabilised sub-bases below that of Type 1 sub-base, the results of the full-scale assessment of the sub-bases were used to estimate thicknesses of the stabilised soil sub-bases and a Type 1 sub-base that behave in a similar manner. The procedure adopted was to establish the sub-base thicknesses that produced the same foundation stiffness. This analysis was carried out on foundations with their stabilised soil sub-bases in a cracked state. Deformation of these equivalent foundations by traffic was then considered to confirm that the selected designs of the stabilised foundations would not deform any more than a traditional unbound foundation.

Equivalent thicknesses of the sub-bases deduced from Figure 9 for a foundation stiffness of 50 MPa are given in Table 7. These thicknesses are specific to foundations built on a soil subgrade of CBR value between 2.5 and 3.0 per cent. The regression lines between foundation stiffness and sub-base thickness used in the analysis were established for the cracked regions of the stabilised sub-bases; that is, the whole of the sub-base CLS-L and the thinnest one third of the sub-base CLS-H. A higher foundation stiffness criterion than 50 MPa to represent a higher quality foundation could have been adopted. It would have allowed equivalencies to be established between the thicknesses of the stabilised sub-bases but would not have permitted comparison with Type 1 sub-base as the maximum thickness of this material in this trial was only about 425mm.

Table 7   Equivalent thicknesses of sub-bases

Sub-base	Equivalent thickness[*] (mm)
CLS-H	225
CLS-L	345
Type 1	405

[[*] Thicknesses rounded to the nearest 5mm.]

The resistance to deformation by traffic of these equivalent foundations was assessed using Figure 10. In this figure, foundation deformation after trafficking is shown as a function of sub-base thickness together with the equivalent thicknesses of each sub-base. Of the three equivalent foundations, those built with stabilised soil sub-bases deformed less and therefore performed better than the foundation constructed with Type 1 sub-base. The performance of the foundation constructed with CLS-H sub-base was particularly impressive as this foundation had carried substantially more traffic than the control foundation.

DISCUSSION

No visible cracks were formed in either stabilised sub-base by environmental factors during the first seven days after construction. However, cracks were formed in these sub-bases by trafficking. The crack pattern induced in the weaker stabilised soil, CLS-L, by permissible wheel loads was considered normal for this material. The measured foundation stiffness was therefore taken to be representative of foundations constructed with this material and to the design adopted in these trials. However, the nature of the cracks within the stronger stabilised soil, CLS-H, was caused by abnormally high wheel loads and may have been different from the crack pattern produced in roads by normal environmental and traffic conditions. Therefore, road trials are recommended for the stronger stabilised soil sub-base to assess whether the values of foundation stiffness measured in the full-scale trials are representative of determinations of foundation stiffness *in situ* .

The trials in the PTF were protected from wet weather. Therefore experiments on the road network are required to assess the effect of changes in weather on the performance of the stabilised soil sub-bases. In particular, investigations are required to determine whether the increased strength of the soil at the top of subgrade, which occurred on construction of the stabilised sub-bases, would be maintained in wet weather.

The full-scale trials demonstrated that lime and cement treated soil can be laid thinner than unbound granular, Type 1 sub-base for a similar structural performance. The recommended thicknesses of sub-base decrease with increase in the strength of the stabilised soil. Alternatively, the stabilised soil could be laid thicker than the derived equivalent thicknesses to provide a foundation of stiffness higher than 50 MPa as shown in Figure 9. This improved support to the pavement would result in an increased pavement life. However, French designs described by SETRA (1981) specify thinner pavement layers, the stiffer the underlying foundation. If a similar approach is adopted in the UK, then economies in pavement construction costs could be made.

The foundation designs given in Table 7 are appropriate to a subgrade of CBR value between 2.5 and 3.0 per cent. For stronger subgrades, thinner sub-bases should be permissible. There is, however, a minimum practical sub-base thickness of 150 mm. In this case, the advantage in using stabilised soil sub-bases of superior stiffness and strength than Type 1 sub-base would be realised either from an increased pavement life, or by a reduced pavement thickness.

Road trials are needed to develop foundation designs for stronger subgrades than CBR 3.0 per cent and to assess reductions in pavement thickness permissible for stiff foundations.

CONCLUSIONS

1. A cohesive soil was successfully treated with lime and cement to produce low and high strength materials, CLS-L and CLS-H, with a CBR value of 35 per cent at 7 days and a compressive cube strength of 1.7 MPa at 7 days respectively. The strength of these materials increased by at least 50 per cent between 7 and 91 days after compaction.

2. On the basis of the satisfactory structural performance of lime and cement treated soil in full-scale trials, it is recommended that these materials should be included in Department of Transport Standards.

3. Foundations of similar structural performance should be produced when sub-bases are laid on a subgrade of CBR 2.5 to 3 per cent and to the following thicknesses: CLS-H, 225mm; CLS-L, 350mm and Type 1, 400mm. The foundations should be more robust than those given in HD 25 (DMRB 7.2.2.) and suitable for heavy traffic during road construction. Alternatively, the stabilised soil sub-bases could be laid thicker than these equivalent thicknesses and the improved support to the pavement used to either increase pavement life or to reduce the thickness of the roadbase layer.

4. Trials on the road network are recommended to establish the effects of typical crack patterns and environmental factors *in situ* . These trials would also be useful in further developing material specifications and designs for the foundation and overlying pavement.

ACKNOWLEDGEMENTS

The work described in this report was carried out in the Civil Engineering Resource Centre (Resource Centre Manager: Mr P G Jordan) of TRL. The assistance of Mr R Addis, Mr S Amor, Mrs V Atkinson, Mr D Blackman and Mr A Halliday of TRL is gratefully acknowledged. The research forms part of the LINK Transport Infrastructure and Operations Programme and is supported by the British Cement Association, Buxton Lime Industries Ltd., County Surveyors' Society and the Department of Transport. The interest and valuable contribution of representatives of these organisations and also Mr J Kennedy and Mr D York is gratefully acknowledged. Advice given by the contractors Powerbetter Developments Ltd and by MSTS Ltd is appreciated. Gratitude is also due to Bomag Ltd for the provision of compaction equipment.

REFERENCES

Black, W P M and Lister, N W (1979), "Strength of Clay Fill Subgrade: its Prediction in Relation to Road Performance", Department of Transport TRRL Laboratory Report LR 889, Transport and Road Research Laboratory, Crowthorne.

British Standards Institution (1980), "Recommendations for Testing Aggregates. Part 1: Compactibility Test for Graded Aggregates", British Standard BS 5835, British Standards Institution, London.

British Standards Institution (1990), "Stabilised Materials for Civil Engineering Purposes. Part 2: Methods of Test for Cement-Stabilised and Lime-Stabilised Materials", British Standard BS 1924, British Standards Institution, London.

Buxton Lime Industries (1990), "Lime Stabilisation Manual", 2nd Edition.

Chaddock, B C J (1988), "Deformation of Road Foundations with Geogrid Reinforcement", Department of Transport TRRL Research Report RR140, Transport and Road Research Laboratory, Crowthorne.

Chaddock, B C J and Brown, A J (1995), "In-situ tests for Road Foundation Assessment", Unbound Aggregates in Roads, Proc. UNBAR 4, Ed. R.H. Jones and A. R. Dawson, Nottingham University, p.259-270.

Chaddock, B C J, Coyle, T and Earland, M G (1995), "The Performance of Recycled Bituminous Bound Materials as Unbound Granular Sub-base", Unbound Aggregates in Roads, Proc. UNBAR 4, Ed. R. H. Jones and A. R. Dawson, Nottingham University, p.193-204.

Dumbleton, M J (1962), "Investigations to Assess the Potentialities of Lime for Soil Stabilization in the United Kingdom", Road Research Technical Paper No 64, HMSO, London.

HA 74 (1995), "Design and Construction of Lime Stabilised Capping", Design Manual forRoads and Bridges, Vol 4, Section 1, Part 6 (DMRB 4.1.6.), HMSO, London.

HD 25 (1994), "Structural Design of New Road Pavements", Design Manual for Roads and Bridges, Vol 7, Section 2, Part 2 (DMRB 7.2.2), HMSO, London.

MCHW 1 (1991) (Manual of Contract Documents for Highway Works), Vol 1, Specification for Highway Works, Amended August 1993, HMSO, London.

Powell, W D, Potter, J F, Mayhew, H C and Nunn, M E (1984), "The Structural Design of Bituminous Roads", Department of Transport TRRL Report LR 1132, Transport Research Laboratory, Crowthorne.

SETRA (1981), "Catalogue des Structures Types de Chaussees Neuves", (Ref. Circulaire R/IN 01/77-1156), Planche Complementaire No 18, Structures en Grave-bitume et en Grave non traitee, Laboratoire Central des Ponts et Chaussees (LCPC), Paris.

Sherwood, P (1993), "Soil Stabilisation with Cement and Lime", HMSO, London.

Sorensen, A and Hayven, M (1982), "The Dynatest 8000 Falling Weight Deflectometer Test System", Proceedings of the International Symposium on Bearing Capacity of Roads and Airfields, Trondheim, 23-25 June 1982, p.464-470.

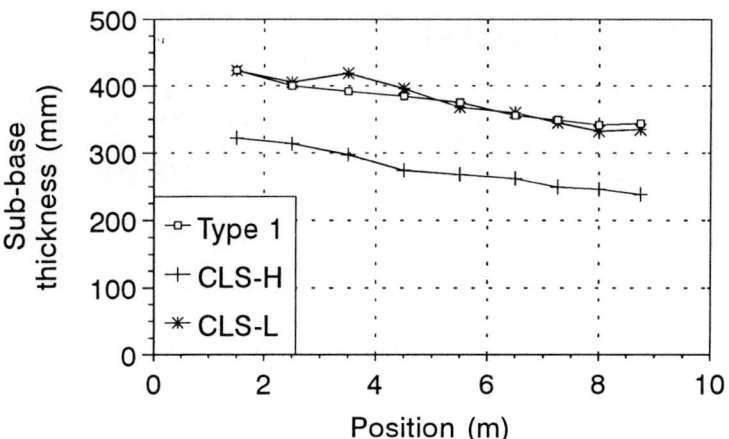

Figure 1 Sub-base thickness along the experimental sections

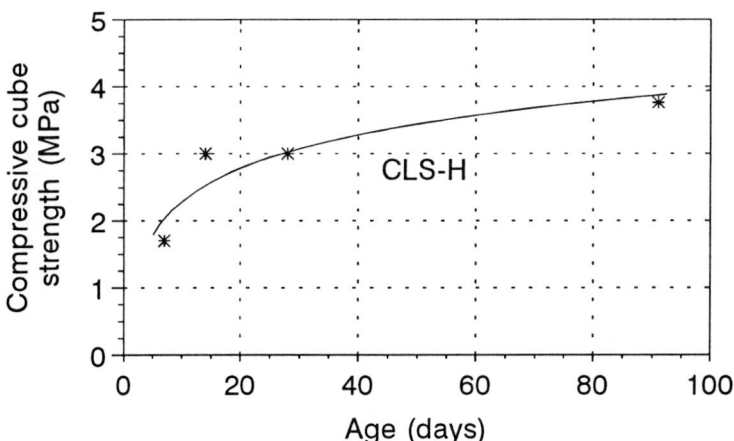

Figure 2 Dependence of compressive strength of stabilised soil sub-base on age

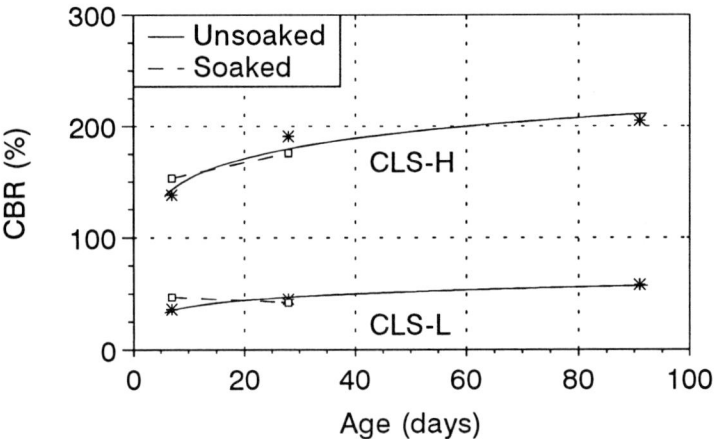

Figure 3   Dependence of CBR of stabilised soil sub-bases on age

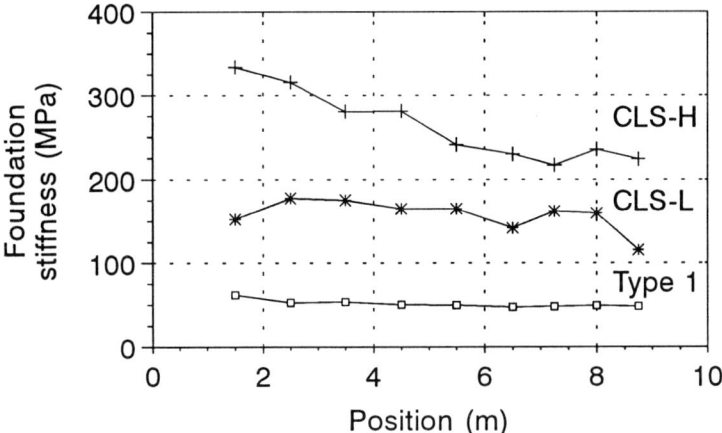

Figure 4   Stiffness of intact foundations along the experimental sections

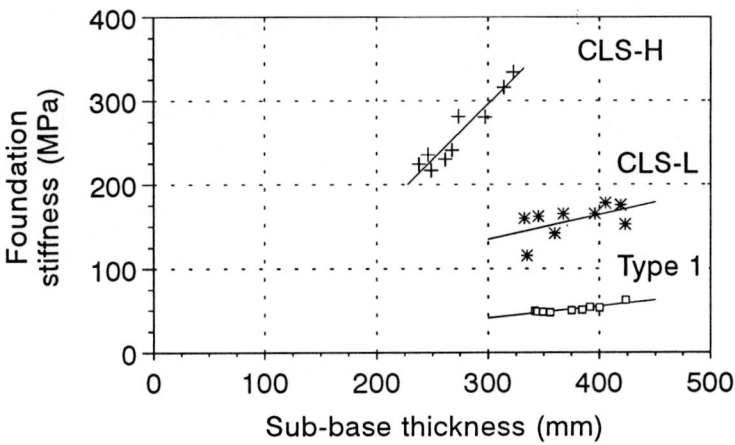

Figure 5　Variation of stiffness of intact foundations with sub-base thickness

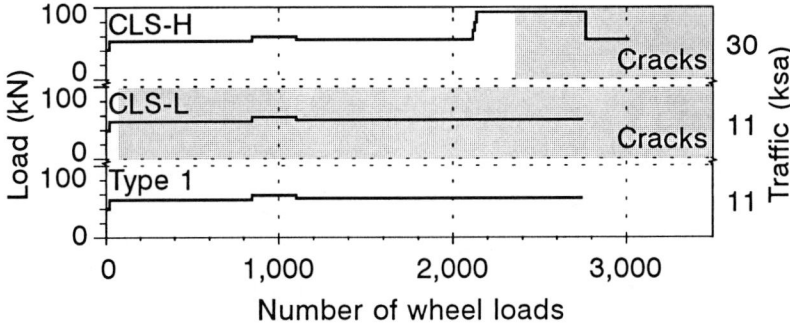

Figure 6　Loading schedule

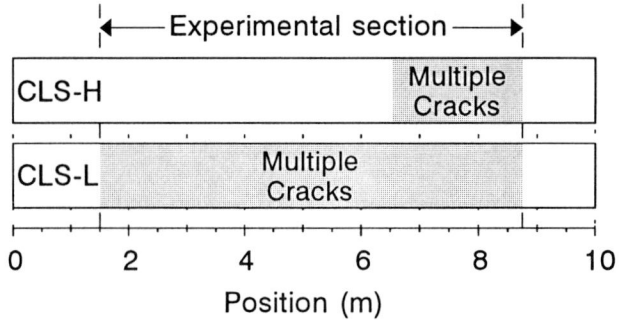

Figure 7   Cracking of stabilised sub-bases

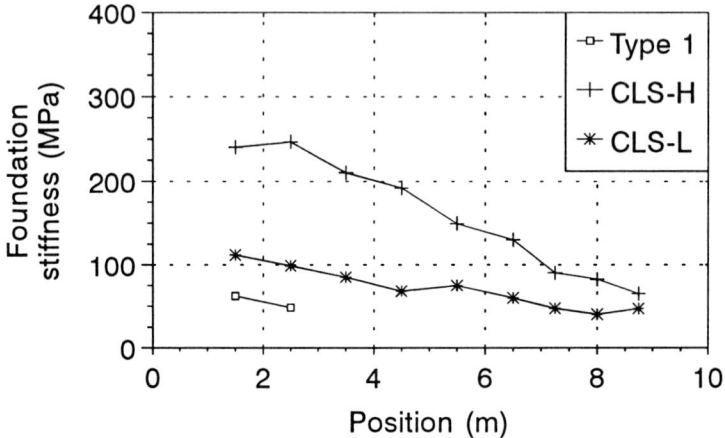

Figure 8   Stiffness of cracked foundations along the experimental sections

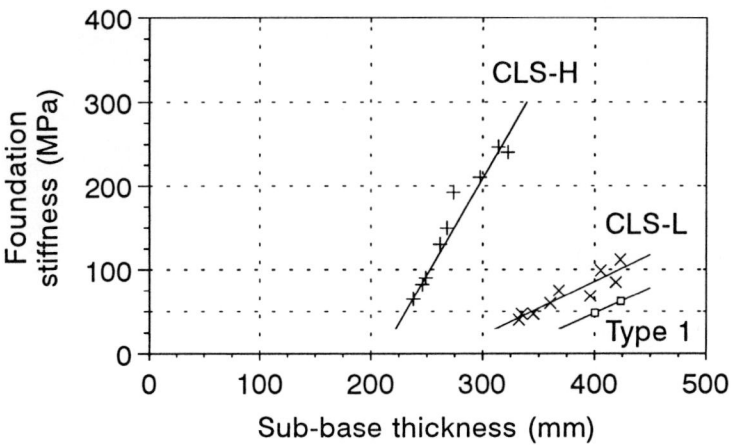

Figure 9   Variation of stiffness of cracked foundation with sub-base thickness

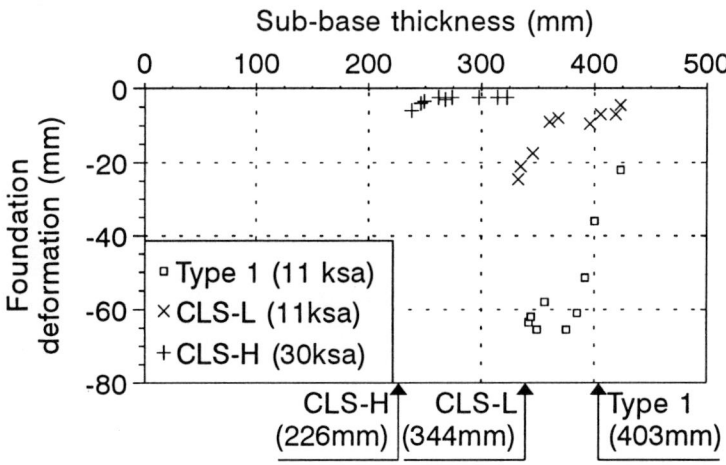

Figure 10   Variation of foundation deformation with sub-base thickness and traffic

NOVEL APPLICATIONS FOR LIME STABILISATION

Introduction

Modification of Clay Soils Using Lime

C D F Rogers Eur Ing, BSc (Hons), PhD, CEng, MICE, MIHT.
Senior Lecturer in Geotechnics, Loughborough University.

S Glendinning BSc (Hons), PhD.
Lecturer in Geotechnics, Loughborough University.

Lime Treatment of Contaminated Sludges

D J Boardman BEng (Hons).
Research Student, Loughborough University.

J A Maclean BSc, (Hons).
Director of Land and Water Services.

Deep Stabilisation Using Lime

S Glendinning BSc (Hons), PhD.
Lecturer in Geotechnics, Loughborough University.

C D F Rogers Eur Ing, BSc (Hons), PhD, CEng, MICE, MIHT.
Senior Lecturer in Geotechnics, Loughborough University.

Introduction

The previous two chapters have discussed lime stabilisation essentially in the context of subgrade and sub-base improvement for pavement construction. The techniques are reasonably well established, enabling detailed information on construction techniques and design specifications to be given. Although there is still much to be learnt about refinement of construction procedures for different applications and details of reaction mechanisms, the basic idea of lime-clay mix improvement is established in the UK construction industry.

The aim of this chapter is to introduce lime stabilisation techniques which are, as yet, in their infancy in this country. The papers provide research data and practical examples of the use of these novel techniques, drawing upon overseas experience where appropriate. It is hoped that these papers will provide sufficient information to allow the reader to asses the viability of these techniques in the context of their own experiences. Ultimately, this chapter may serve as an impetus for the greater acceptance of novel techniques and provide some confidence for potential users that the techniques can prove both successful and cost efficient.

The first paper examines the role of modification of clay soils using lime. The reader is reminded of the two-phase reaction brought about when lime is mixed with clay soil. The first phase is immediate and brings about changes in plasticity, strength and water content. It is termed modification. Only in the long term does stabilisation occur through chemical reaction to bring about very large improvements in strength. This paper examines the goals of lime stabilisation: what does the engineer require of the improved material? It is pointed out that if what is required is a rapid drying out and reduction in plasticity for, say, bulk filling during inclement weather, then modification is the goal for the process. As has been shown previously (Holt and Freer-Hewish in Chapter 2) much smaller quantities of lime are required to produce modification than stabilisation. Current practice, and the abiquitously conservative approach of the construction industry, lead to the application of more than sufficient lime to fulfil stabilisation requirements. If the ideas discussed in this paper were put into practice, considerable cost saving could be expected. The paper goes on to discuss the practical considerations of mixing and compaction processes and provides an example of its usage overseas. This should provide food-for-thought for an industry troubled by the costs of wet winter working.

The second paper addresses the extremely topical subject of clean-up of contaminated wastes. It offers lime treatment as a potentially efficient and cost effective solution. The legacy of industrial wastes is revealed. With industry producing increasing quantities of hazardous materials and legislation requiring constantly increasing clean-up standards, a reliable and cost effective treatment process is required.

This paper aims to introduce the reader to the concept of solidification and stabilisation treatments for a variety of industrial waste sludges. The complexity of the sludges and their possible treatments is evident, although the paper aims to simplify the discussion where possible. The theory behind lime treatment is explained and the development of the first UK treatment system presented. It is explained that the complexity and variability of waste sludges means that the system cannot provide a *carte blanche* treatment for all situations. However, the UK treatment system is very flexible and can offer a wide-ranging applicability. It appears to be a potentially exciting development for an ever-increasing problem, but clearly considerable further research is required before questions about applicability, reaction mechanisms and construction techniques can be answered. Considerable investment and courage from the construction industry in a new technique is the ultimate answer.

The third and final paper explores deep lime stabilisation techniques, as opposed to the surface mixing processes already presented. Potential applications are seen to be stabilisation *in situ* of soft soils beneath new transportation routes or structures, or remedial treatment *in situ* for failed cuttings or embankment slopes.

Three techniques are discussed: lime columns; lime slurry pressure injection and lime piles. These have previously been discussed by Perry *et al*, in Chapter 1. Lime columns are mixed-in-place lime stabilised units which have been used primarily in Scandinavia and Japan for the treatment of soft soils. The lime-clay reaction is employed to produce columns of stronger, stiffer material than the surrounding soil. Specialist equipment is required to produce the columns, requiring considerable investment from specialist contractors.

Lime slurry pressure injection has been used primarily in the USA for slope stabilisation. Again, investment is required in the 'injection' equipment and some doubt has been expressed about its reliability.

The research at Loughborough University into lime piles is described. Lime piles are essentially columns of lime alone, requiring no mixing and therefore no specialist equipment. Their use overseas is explained, although UK experience is concentrated upon, where the piles are used for slope stability. The mechanism of stabilisation is presented from data gained from both laboratory and field applications. The first UK commercial application is described by Threadgold in the next chapter.

Clearly, these are examples of techniques new to the UK and there is much work to be done before they can become well-established. It is the responsibility of academia and industry to make partnerships to explore, concurrently, the issues of technical detail and reliability, construction processes and economics.

Modification of Clay Soils Using Lime

C.D.F. ROGERS AND S. GLENDINNING
Civil and Building Engineering Department, Loughborough University

INTRODUCTION

Lime has been used for centuries to improve soils for construction. Most notably the Romans used lime to strength and stiffen clay soils. Lime stabilisation has traditionally been far more widely used overseas than in the UK. There are several possible reasons for this including the historical approach to engineering works in the UK, the availability of local sources of crushed rock and the conservatism of engineering practice embodied by, for example, recipe specifications for road pavement construction. A more important reason, however, is likely to be the perception that lime stabilisation requires a hotter climate than that in the UK to be successful. While this is emphatically not the case, the perception was reinforced by its extensive use in countries having warmer climates. Further reinforcement perhaps derives from the idea that the goal of lime stabilisation is solely to create a very strong, stiff cemented material that can act in the same manner as a weak concrete slab and that high temperatures are required to achieve this in the minimum possible time. In fact lime has considerable potential for use in UK construction to achieve several different goals, but in order to be used effectively it must be incorporated into a design that has the specific benefits sought clearly isolated.

The creation of full lime stabilisation requires a significant percentage of lime to be added to and mixed with the clay, an adequate understanding of the reaction processes and a good knowledge of the compaction process. It thus requires careful design and close attention to detail during the construction process in order to ensure that the long-term benefits are achieved. Addition of much smaller quantities of lime, however, can produce considerable benefits both in the long term and during construction. One obvious benefit derives from the hydration of quicklime causing drying of clay soils on site. A more important process is that of modification of a clay soil by the addition of lime, which changes its fundamental nature to create a material that is more workable (i.e. less plastic) and stronger.

The aim of this paper is to examine the changes wrought on different clay soils by the addition of relatively small quantities of lime. In particular the different effects that occur with different clay types will be highlighted and the quantities of lime required to achieve these effects in each case will be discussed. Finally the methods of construction will be considered in the light of the recommendations made from laboratory test data.

LIME CLAY REACTIONS

Introduction

The reaction between lime and clay can be divided into two distinct processes: modification and stabilisation (Sherwood, 1993). Modification occurs rapidly after addition and mixing of lime with a clay. This occurs typically within 24 hours, although it sometimes takes up to 72 hours depending on the clay minerals involved. Stabilisation refers to a process that occurs far more slowly, producing a long-term strength gain by the progressive crystallisation of gels that are created once the lime has attacked the clay minerals. This paper is concerned with modification, which is discussed in greater detail below.

Lime Modification

When quicklime (calcium oxide) is added to a clay soil the following dehydration reaction takes place to remove water from the clay:

$$CaO + H_2O \rightarrow Ca(OH)_2 + heat \qquad\qquad\qquad Eqn\ 1$$

Since the reaction is strongly exothermic (65kJ/mol is generated), additional water will be driven off through the production of steam if the reaction takes place quickly enough. The strength and workability of the clay is improved merely by the drying action of lime.

In addition, calcium ions released by the lime may exchange with metal ions that are associated with the clay lattice by the process of cation exchange. The ability of a clay to undergo this change is measured by its Cation Exchange Capacity (CEC). Observed changes and their effects on engineering properties include the following:

- A substantial reduction in the thickness, and thereby stabilisation, of the adsorbed water layer (known as the electrical double layer) occurs. This results in a reduced susceptibility of the clay to the addition of water.

- Flocculation of the clay particles takes place. This is caused by greater attraction between the clay particles that occurs due to their closer proximity, which in turn is a result of the reduction in the thickness of the water layer.

- There is an increased internal angle of friction between the agglomerates, and hence greater aggregate shear strength under any given normal stress conditions.

- A textural change takes place from a plastic clay to a friable material that is granular in nature.

- The clay exhibits reduced plasticity, as evidenced by a significant reduction in the Plasticity Index that is usually caused by a considerable increase in the Plastic Limit of the clay.

These changes are clearly visible in the laboratory or on site since the changes to the nature of the clay are dramatic and almost instantaneous on mixing with lime. The progress of the reaction is, however, best monitored by carrying out Atterberg Limit tests, the reaction being complete once the values of Liquid and Plastic Limit (LL and PL respectively) have stabilised.

The factors that affect this process are related to the clay itself and the efficiency with which mixing can take place. Clay minerals with a high CEC, such as montmorillonite, undergo a large change in properties on mixing with lime whereas those with a low CEC, such as illite, undergo a small change. A natural clay will typically consist of a number of clay minerals, which complicates the matter. The degree to which mixing can take place is determined by such factors as the water content of the clay, the distribution and fineness of the lime particles, the type of mixing plant available and the size of the clay lumps (or degree of pulverisation) that is achieved by the plant on the particular clay. These factors together, in turn, will determine the rate at which the modification reactions take place since they control the speed with which cation exchange conditions (i.e. the concentration of calcium ions) can be created throughout the clay mass.

Lime Stabilisation

To achieve full stabilisation additional requirements must be met. Firstly the quantity of lime added to the clay must be sufficient to produce full modification of the clay minerals (this quantity is known as the lime fixation percentage addition) *and* thereafter to cause the stabilisation reactions to occur. The percentage of lime greater than that needed to achieve lime fixation is generally determined by laboratory testing. The lime must then be intimately mixed with the clay, rolled to create a seal against excessive attack by carbon dioxide in the air and left to mellow for sufficient time for modification to take place fully. The modified clay must then be mixed with the correct amount of water to achieve a water content approximately equal to or slightly greater than the optimum water content and it must then be thoroughly compacted to minimise air voids and maximise density.

It is in this dense state that the stabilisation reactions can take place to produce a strong, stiff cemented material. The hydroxyl ions released from the lime create a pH level that is sufficiently high that silica and alumina are dissolved from the clay minerals, from which new compounds are formed as a result of pozzolanic reactions. The silica and alumina within the clay structure react with the water and lime to form calcium silicate hydrate and calcium alluminate hydrate gels, which subsequently crystallise to bind the structure together. As a result of these reactions the material becomes stronger and more brittle, the progress of the reactions being determined by conventional shear strength tests.

EVIDENCE FROM THE LITERATURE

Introduction

As mentioned earlier, the evidence of lime modification and stabilisation having taken place in a lime-treated clay is via the three primary factors of observations of material nature, change in plasticity and increase in strength. The latter two parameters are commonly measured in practice, although it is the former that provides the most reliable evidence of modification and it is therefore plasticity that will provide the focus in this paper.

It should be noted that the changes sought on site with respect to facilitating construction works by lime addition are generally related to the workability of a clay soil. The first requirement would be to address the lack of trafficability on a soft, wet clay, for which the dehydration effects of quicklime would provide immediate assistance. The second goal of mixing quicklime into the surface layer would be to increase the PL of the clay. For example if a clay with a water content of 30% has its PL raised from 24% to 32%, then it will become non-plastic by virtue of modification and hence trafficable. The same arguments can be applied to material that is to be used as cut and fill. In such cases as these the requirement is not one of full modification necessarily, but of sufficient change to facilitate working. Thus the percentage of lime that needs to be mixed effectively with the soil will be consequently lower than the lime fixation value. This idea appears to have received little attention in the literature as it is full stabilisation that is commonly tested for. Nevertheless it is apparent that the plasticity changes, and in many cases the strength gains, quoted for lime contents at or below the fixation levels (where tests have been performed over a range of lime additions) are sufficient for many applications.

One further point of importance is that the method of determining the point of lime fixation will not always result in a consistent measurement. The method currently used is the Initial Consumption of Lime (ICL) test (BSI, 1990), in which the measurement of an absolute level of pH is used to determine fixation. However variation in the method of measuring pH, the freshness of the lime used and possibly also the clay mineral types might all affect the result. The current authors advocate consideration of the full pH - lime addition curve (full modification occurring as the asymptote is approached), which will in some cases result in lower fixation levels than those quoted. This argument is supported by Thompson and Eades

(1970), who comment on the Quick Test developed by Eades and Grim (1966) which formed the basis of the ICL test. Using measurements of compressive strength after 28 days, they found that in 16 out of 27 cases the Quick Test overestimated the amount of lime required by 1 to 2% with an average overestimation of 0.74%. After 56 days overestimation occurred in 24 out of 26 cases with an average of 0.54% lime addition. This is discussed in more detail by Rogers et al (1996), but is included herein since it might affect the interpretation of data reported in the literature.

In addition, Dumbleton (1962) demonstrated that the effects of lime on plasticity vary with clay type, percentage lime addition and time. It is important therefore that the testing procedures are examined closely to ensure that sufficient time has been allowed for the modification reactions to take place when considering the results reported in the literature, but that long-term effects do not mask the results.

Tests on British Clays

Dumbleton (1962) reports the results of a series of tests on a range of UK clays. In the case of plasticity changes, the only detailed findings presented are for London Clay in the form of two graphs. The clay was mixed with hydrated lime and stored in air-tight jars at water contents in the region of the initial LL and PL values. The temperature at which the samples were stored was not recorded. Atterberg Limit tests were subsequently carried out on the samples at various intervals of curing up to one year. Data extracted from the two graphs are presented in Table 1.

Table 1. Change in plasticity of London Clay caused by lime addition (after Dumbleton, 1962)

Lime Addition (%)	Plastic Limit (%)			Liquid Limit (%)			Plasticity Index (%)		
	Imm	1 mth	1 yr	Imm	1 mth	1 yr	Imm	1 mth	1 yr
0	26	28	27	80	77	86	54	49	59
1	34	35	32	93	82	84	59	47	52
2	43	44	33	87	89	92	44	45	59
4	46	64	62	79	96	102	33	32	40
7	48	62	85	81	90	117	33	28	32
10	49	68	90	76	92	112	27	24	22

Notes: 1. Measurements were taken immediately (Imm), after one month and after one year.

It is clear from the immediate PL data that a considerable increase occurred with the addition of 1% lime, and that a similarly large increase was effected by a further 1% addition. Once 4% lime has been added, however, very little further change was apparent and thus it might be concluded that the fixation level for London Clay lies between 2% and 4% lime addition. This is confirmed by the data at one month, which show similar changes in PL for 1 and 2% lime addition but an increase of 14-19% for 4-10% lime addition when compared with the immediate PL data. These data indicate that the (latter) material tested had undergone stabilisation reactions, causing a further fundamental change in the nature of the material. Dumbleton states that, where necessary, the material was broken up by a pestle and mortar prior to testing, thus confirming that full stabilisation had occurred in some cases. The PL data for one year are likewise consistent, with additional stabilisation occurring for 7 and 10% lime addition. The reduction in PL for 2% lime addition after one year is not discussed and, although some marginal lowering could be discerned for the 6 month data the value should not be considered a significant finding without further testing, particularly since PL test data on stabilised soil can be difficult to interpret precisely. Overall the PL trends are consistent with expectations.

LL determinations traditionally produce greater variation in observed effects since LL is far more sensitive than PL to the cation present (Diamond and Kinter, 1965). The data given in Table 1 show an immediate increase for the lower lime additions, with little change for 4% or more lime. After longer periods there appears to be negligible change for 1% lime, some change for 2% lime and a generally greater increase for 4-7% lime. These results thus broadly confirm the observations from the PL data.

A clearer distinction is given by the plasticity index (PI) data, which show a clear drop for lime additions of 4-10% when compared with those of 1 and 2%. The immediate increase in PI with 1% lime is caused by the increase in LL. This is a commonly reported phenomenon for London Clay, but since it is the PL that is most important for site work the PI increase has little relevance. It is clear, therefore, that emphasis should be placed on the PL data for determining workability of materials on site, and hence modification requirements.

Further data on London Clay plasticity changes are presented by Sherwood (1993) and Bell and Coulthard (1990). Since the values appear to be identical, only one set has been reproduced here (Table 2). The trends in the values replicate those of Dumbleton, hence confirming the observation that the fixation point lies between 2 and 4% lime addition.

Table 2. Change in plasticity of London Clay caused by lime addition (after Sherwood, 1993).

Lime Addition (%)	Plastic Limit (%)	Liquid Limit (%)	Plasticity Index (%)
0	24	80	56
2	40	84	44
4	44	77	33
7	45	80	35
10	45	76	31

Cobbe (1981) reports plasticity changes wrought on three Jurassic clays and these are shown in Table 3. In each case the clay was dried and passed through a 425μm sieve prior to testing. For the PL determination the clay powder was mixed with lime, water was added and the mixture was left to equilibrate for one hour prior to testing. For the LL determination the clay powder was wetted and allowed to equilibrate for 24 hours prior to mixing with lime and being allowed to cure for a further hour prior to testing. The PL data for all three clays show a remarkably consistent trend of rising to an asymptote. The LL data produce a similar, though less well defined, trend with significantly higher variability at the higher lime contents in each case. This results in a rising trend of PI, with overall increases of approximately 15 for the silty clays and 30 for the highly plastic clay.

For Upper Lias Clay the change in PL and LL is almost complete with 1.5% lime and the fixation point could be deduced to be reached. For the Oxford Clay having a similar plasticity, 1.5% lime again produces most of the change seen by higher lime contents, but an addition of 2.25% was needed to achieve the full change and thus the fixation point will be between the two. For the heavier Blisworth Clay the PL data indicate a fixation point around 3.0%, with the LL data appearing to stabilise earlier.

While the point of full change is of importance, it is clear that the changes produced by the lower lime addition in each case are substantial. The changes in PL alone are sufficient to make wet sites and wet material potentially workable. While further data are clearly required to establish the changes caused by small lime additions, the practicality of mixing small quantities of lime efficiently and uniformly also need to be addressed.

Table 3. Summary of plasticity changes to three Jurassic Clays caused by lime addition (after Cobbe, 1981).

Clay Type	Lime Content (%)	Plastic Limit (%)	Liquid Limit (%)	Plasticity Index (%)	Mineralogy
Upper Lias Clay	0	24	53	29	K - abundant
	1.5	38	80	42	
	2.25	39	86	47	I - common
	3.0	40	83	43	
	4.5	42	84	42	C - trace
	6.0	41	87	46	
Blisworth Lay	0	32	72	40	K - common
	1.0	39	108	69	
	2.0	46	118	72	I - common/
	3.0	49	124	75	abundant
	4.0	52	114	62	
	6.0	51	112	61	C - trace
	8.0	53	120	67	
Oxford Clay	0	26	54	28	K - common
	1.5	37	73	36	
	2.25	40	81	41	I - abundant
	3.0	40	77	37	
	4.5	42	84	42	
	6.0	43	85	42	
	8.0	44	84	40	

Note: 1. K = kaolinite, I = illite, c = chlorite.

Overseas Research

Although much work has been done on the modification of clay soils worldwide, two examples only are reported here to illustrate specifically the effects of time after mixing and efficiency of mixing small proportions of lime with clay on site. Osula (1996) reported a study of Nigerian laterite modification to examine the practicality of use in road construction. The laterite was a highly weathered tropical soil having a clay content of 13% consisting predominantly of kaolinite. The laterite was mixed thoroughly with lime in the presence of water in a tray. It was thereafter tested either immediately or allowed to mellow for periods of 1, 2 or 3 hours prior to testing.

Plasticity data extracted from the graphs presented by Osula are presented in Table 4. It is apparent that a lime addition of between 1 and 2% is necessary to modify this material and equally that full benefits are not achieved immediately after mixing. This is to be expected, unless perhaps the lime and the soil are to be intimately mixed in a powdered form prior to wetting, since cation exchange needs to take place fully within the clay lumps. The data show that modification takes place quicker with the higher lime contents, full change occurring after 2 hours with 3% lime and after approximately 3 hours with 2% lime. It is unclear whether a further change in plasticity with 1% lime addition would occur beyond 3 hours since no further data are available. However, since the fall in LL is consistent over the three hour period further reduction would appear likely.

The changes in the properties of the laterite are considerably greater to the LL than the PL, to the extent that it is the considerable reduction in LL that makes the material almost non-plastic after full modification. It is, however, the influence of time after mixing that is most important in this case. Mixing lime with clay by hand in the laboratory ensures a uniform distribution around relatively small clay lumps, and thus the full effects of relatively small lime contents can be realised. In the field, however, this is perhaps more difficult to ensure and some compensation should be allowed in terms of an increased curing period to allow cation exchange to take place fully.

This practically has been examined by Sweeney (1987), who studied the lime treatment of two clays of high plasticity for road foundation construction in Canada. Disturbed samples of the clay beneath Highway 339 were taken and mixed with 1, 2 and 3% lime in the laboratory to provide control sample data, as shown in Table 5. Measurements of pH were also recorded. Samples were also taken following site mixing from sites on Highway 339, where 2.5% lime was added, and Highway 334, where 1.0% lime treatment was used. These samples were similarly tested for Atterberg Limits and pH (Table 5).

Table 5. Plasticity changes to Canadian clays beneath Highways 339 and 334 (after Sweeney, 1987).

Clay Sample Tested	Lime Addition (%)	Plastic Limit (%)	Liquid Limit (%)	Plasticity Index (%)	pH
Laboratory Samples (Highway 339)	0	22	72	50	7.8
	1	29	60	31	10.4
	2	38	55	17	11.6
	3	43	56	13	11.9
Highway 339:					
Untreated	0	24	78	54	7.9
	0	23	78	55	7.8
Bottom lift	2.5	30	56	26	11.9
	2.5	31	62	31	10.8
	2.5R	36	59	23	10.7
Middle lift	2.5			NP	12.2
	2.5	39	58	19	10.8
	2.5R	34	56	22	10.4
	2.5R			NP	12.1
Highway 334:					
Untreated	0	21	63	42	8.3
Top lift (A)	1.0	23	40	17	9.8
	1.0R	23	38	15	10.0
Bottom lift	1.0	31	67	36	10.7
	1.0	34	64	30	11.1
	1.0R	29	65	36	10.6
	1.0R	34	68	34	10.8
Top lift (B)	1.0	30	66	36	9.6
	1.0R	30	63	33	10.7

Notes: 1. R against the lime addition indicates that plasticity was measured after remixing.

2. The top lift of the stabilised layers was tested at two different locations (termed A and B) along Highway 334.

3. NP indicates that the sample was non-plastic.

The clay preparation procedure adopted on Highway 339 was pulverisation using a disc harrow, graders fitted with rippers and two "pulvimixers" to achieve a clay lump size of generally less than 100mm. The material was then treated in two ways. In the case of the northbound carriageway lime was spread on the top 150mm lift, "dry mixing" took place and the material was stockpiled at the edge of the road to mellow. The middle 150mm lift was treated in the same way. Lime was spread on the bottom 150mm lift once pulverised and "dry mixing" took place, as above. Water was then added by several passes of a water bowser, followed immediately by compaction without remixing (i.e. without "wet mixing"). This was found in some cases to produce poor distribution of lime. The upper lifts were then sequentially bladed onto the road foundation surface, mixed, levelled, wetted to an appropriate water content and compacted. Compaction was effected using a sheepsfoot roller.

On the southbound carriageway the two upper pulverised soil lifts were stockpiled at the edge of the road without lime being added. The bottom lift was then treated as above, but a second stage of mixing using the rippers and "pulvimixers" was introduced prior to wetting and compaction. The upper two lifts were then sequentially bladed back and treated in the same way. This caused the mellowing period of the upper lifts to reduce from 3-4 days to no longer than one day when compared with the northbound carriageway.

The plasticity data for Highway 339 show that addition of 2.5% lime and "dry mixing" in the base layer did not produce the same reduction in PI as addition of 2.0% lime in the laboratory, although it should be noted that the PI of the untreated sample taken prior to mixing on site was marginally higher than the control. This was primarily because the PL increase was lower in the field, although remixing did appear to improve the situation to give approximate equivalence to 2.0% lime addition in the laboratory. The middle lift, which had additional working and mellowing, produced better results than the lower lift, giving equivalent data to the 2.0% control samples in two cases and non-plastic results in the other two. It is interesting to note that the two non-plastic samples had higher values of pH (12.1-12.2) than those for the 3.0% lime addition in the control sample (11.9). Remixing of this layer had little apparent effect. A graph of pH against PI is presented in Figure 1, in which it can be clearly seen that the results for the middle lift are significantly better.

The PI of the subgrade in the sections treated beneath Highway 334 was significantly lower (42%) than that under Highway 339 (50-55%). The reason for this was that the subgrade beneath this highway had been nominally modified by 2.0% lime in a poorly controlled operation three years earlier, in which certain sections received a lower dosage and were poorly mixed. This resulted in poor performance, as testified by the still high PI values for the untreated material. The previous treatment is the reason why a lower lime addition of 1.0% was used. The top 150mm lift of subgrade was pulverised using a disc harrow and graders fitted with rippers until the clay lumps did not exceed 100mm in size. This was apparently achieved relatively rapidly due to the previous lime treatment. The lime addition of 1% was spread evenly over the clay surface and was "dry mixed" using the rippers and discs, incidentally causing further pulverisation. Water was then added to bring the material to a water content 2-3% above the optimum water content and the material was "wet mixed" using a "pulvimixer". This material was then stockpiled beside the road to mellow for at least one day. The lower 150mm lift was then treated in the same way except that the material was immediately compacted using sheepsfoot rollers following the "wet mixing" (i.e. it had no time to mellow prior to compaction). The top lift was then bladed into place, further water was added if necessary and the material was remixed prior to compaction. The "wet mixing" process apparently produced a much more uniform distribution of lime and a far better modification process.

The plasticity data shown in Table 5 indicate some variability in the results depending upon the location, which was apparently the result of the differentially effective previous treatment. The bottom lift produced consistent data with a significant increase in PL from 21% to 29-34% and a marginal increase in LL producing a relatively small reduction in PI from 42% to 30-36%. The top lift in area B produced similar results, in spite of the additional mellowing period, whereas the top lift in area A produced only a marginal increase in PL but a large reduction in PI to 15-17%. This larger decrease in PI was more typical of the more extensive Department of Highways data, which indicated average PIs (covering both lifts) of 9.3 and 21.1 for Areas A and B respectively. The results are plotted in Figure 2 together with the control data for Highway 339. Although the majority of the results conform to a lime addition marginally lower than the 1.0% control samples, care should be taken in their interpretation since the clay used for the control sample derived from a different site. The results for the upper lift in Area A show an improvement in plasticity of between 2.0 and 3.0% when compared with the control samples, presumably as a result of the combination of the two treatments.

RESEARCH AT LOUGHBOROUGH UNIVERSITY

A programme of research has been conducted at Loughborough to examine the effect of adding lime at contents lower than the ICL value to four British clays. The first point that became apparent was that the British Standard (BSI, 1990) method of measuring the lime fixation point produced values that were generally far higher than those interpreted from plasticity data, as discussed earlier. A modified interpretation using the full pH - lime addition curve was thus proposed by Rogers et al (1996). The ICL values recorded for the standard method for Weathered Mercia Mudstone (formerly known as Keuper Marl), Lower Lias Clay and London Clay were 1.5, 6.0 and 5.0, whereas those for the modified method were 1.0, 3.0 and 3.0 respectively. No value for the ICL of English China Clay (ECC, a refined material from Cornwall) could be obtained using the standard method because the pH level did not rise high enough, whereas the modified interpretation produced a value of 1.5%. The subsequently derived plasticity data confirmed that these values were generally more accurate, although even these levels could be considered to be high.

The clays were dried at 110C for 24 hours, cleaned of organic matter where appropriate and passed through a 425μm sieve. The clay was then intimately mixed with fresh quicklime, which had been heated to 450C for 24 hours prior to use to ensure that the quicklime was as pure as possible. Distilled water was added to produce a soft, wet paste consistency (typically at a Liquidity Index of 50%) and the material was then sealed in heavy duty polythene bags that were overwrapped and stored at the ambient laboratory temperature (20 ± 2°C) until required for testing. The lime-clay mixes were tested for LL and PL at 0, 3, 6, 24, 48 and 72 hours after mixing at the lime contents shown in Table 6.

108

Table 6. Summary of plasticity changes to four British Clays caused by lime addition (after Rogers et al, 1996).

Clay Type	Lime Addition (%)	Plastic Limit (%)	Liquid Limit (%)	Plasticity Index (%)	Mineralogy
English China Clay	0	35	61	26	Pure well crystalline kaolinite
	0.5	39	68	29	
	1.0	40	74	34	
	1.5	41	72	31	
	2.5	39	71	32	
Weathered Mercia Mudstone	0	19	30	11	Illite, dolomite and quartz
	0.2	20	32	12	
	0.4	21	36	15	
	0.6	25	38	13	
	1.0	25	38	13	
Lower Lias Clay	0	23	62	39	Disordered kaolinite, illite and quartz
	1.0	39	85	46	
	2.0	47	85	38	
	3.0	46	85	39	
	6.0	51	79	28	
London Clay	0	19	52	33	Illite, disordered kaolinite and smectite
	1.0	26	66	40	
	2.0	36	66	30	
	3.0	38	67	29	
	7.0	39	65	26	

Using the logic discussed above, it was apparent that full modification occurred (i.e. the lime fixation point was achieved) for ECC at approximately 1.0% lime, for Keuper Marl at 0.4-0.6% lime, for Lower Lias Clay at 2.0% lime and for London Clay at 2.0-3.0% lime. The values used to construct Table 6 were taken at different times after mixing that suggested that full change had occurred, although the caveat that longer-term changes might be occurring at the higher lime contents should be applied here. In all cases the data reported in Table 6 were for changes at 24 or 48 hours. The same patterns of steadily increasing PL and slightly more erratically increasing LL were found from this work, yielding a somewhat variable pattern of PI variation. The significant, and in some cases considerable, changes caused by a low lime percentage addition were also evident.

DISCUSSION

It is evident from the above data that the addition of quicklime to a wet clay soil causes improvement by drying and reduced susceptibility to water by the increased PL reduction in PI. It is equally clear that the threefold manner of the improvement is likely to be (far) more than adequate for most construction operations to continue under inclement environmental conditions. If this is the primary, or sole, reason for adopting lime modification then a rapid means of assessment of required lime addition is possible using the Atterberg Limits. In general it is the increase in PL relative to the water content of the clay that is most important. This is fortunate since the relationship between PL and lime addition is consistent, whereas that for LL (and hence PI) is considerably more variable. Thus in practice the curve of PL against lime addition should be established by carrying out tests at several lime contents between the ICL value (BSI, 1996) and zero addition, with particular emphasis on the lower range of lime additions since the ICL value is likely to be (significantly) conservative. In view of the simplicity of the PL test and the need only for disturbed, bag samples, replicate readings of the PL should be taken to confirm the relationship derived. It is evident that clay mineralogy and clay fraction are the primary governing factors in determining the volume of lime needed to bring about any specific change in plasticity, those clays having a high CEC being most responsive to lime modification.

Having determined the target value required, it is important to consider the practical implications of attempting to modify a clay with a low lime content. The first consideration is whether uniformity of application can be achieved when applying a small percentage of lime. Incorporation of the lime would need to be as thorough as if full stabilisation were being sought and as thin a layer as would be sufficient for the application should be used. The tendency would be to increase the percentage of lime added to offset the consequences of lack of uniformity of mixing, but this might not prove to be necessary and thus would incur unnecessary expense. The effectiveness of lime modification has been shown to improve by allowing the lime-clay mix to mellow for the time required for full cation exchange to occur. Again the percentage of lime could be increased to reach the same target improvement in a shorter period of time. In either of the above cases, however, the target lime content is likely to be low and thus the financial implications of marginal increases are likely to be insubstantial.

A further practical consideration is that for full stabilisation, care is usually taken not to compact a lime-clay mix before the modification reactions are complete. This is because the hydration reaction is expansive and would cause the effects of compaction subsequently to become (partially) lost. It might not prove to be important in the case considered here since the volume of lime is generally small and water is typically available for hydration. Nevertheless the phenomenon should be considered in design. A further consideration is the permanency of the changes wrought by modification, where these are considered necessary or desirable. There appears to be a lack of evidence that the modification reactions are permanent. This is not to suggest that they are not permanent since there is no evidence to demonstrate this either, excepting perhaps the one reading for London Clay quoted by Dumbleton (1962). Thus further research into this aspect is desirable, particularly because the current authors are of the opinion that the civil engineering industry tends to seek full stabilisation in some cases where a (much) lower strength and stiffness would suffice.

CONCLUSIONS

Lime modification has considerable potential in the civil engineering industry to assist with the construction process and to bring about permanent changes to clay materials. Such assistance could prove to have relatively large economic benefits due to the simplicity of the process and the potential, for example, to extend the earthworks season for major construction schemes. It is apparent from both the literature and the research programme at Loughborough University that several indicators to the successful use of lime modification can be discerned. These are as follows:

- Lime modification causes considerable change to the nature of a clay at lime contents below the fixation (i.e. ICL) level. Quicklime will dehydrate a clay, increase its Plastic Limit and typically reduce its plasticity index. The clay will thus become drier and less susceptible to water content changes.

- The changes in plasticity are governed mainly by the mineralogy of the clay and the proportion of clay contained within the soil (i.e. the clay fraction).

- The amount of lime required fully to modify a clay soil is best determined by the modified ICL test, in which the full pH-lime addition curve is used in interpretation, or by reference to plasticity changes.

- The Plastic Limit (PL) is the best indicator of the lime content necessary to achieve the degree of modification sought in general since the pattern of PL change is consistent for any one clay. Liquid Limit, and hence plasticity index, tends to produce a more variable pattern of changes.

- Efficient mixing to produce a uniform lime addition is necessary to minimise the quantity of lime required to achieve the improvements sought.

- The effect of lime modification increases with time after mixing until the point which full cation exchange (for that lime addition) is achieved. For this reason as long a mellowing period (up to the time required for full cation exchange) as is practical should be used. The period to achieve a target effect can be shortened by increasing the quantity of lime added. Only marginal increases should be necessary.

- The quantity of lime required *to achieve the degree of improvement sought* should be determined. The tendency has been to add larger quantities of lime and seek benefits greater than those needed.

- The long-term effects of small lime additions on the properties of clay soil appear not to have been researched. This should be addressed, as should be the need for more short-term data on British Clays, in order that better use of the technique can be made in industry.

REFERENCES

British Standards Institution (1990), "Methods of Test for Stabilised Soils", BS1924, HMSO, London.

Cobbe, M I (1981), "A Preliminary Investigation into Lime Modification of Heavy Clays", Postgraduate Diploma Dissertation, Middlesex Polytechnic.

Diamond, W and Kinter, B (1965), "Mechanisms of Soil-Lime Stabilisation", Highway Research Record No. 92, TRB, Washington D.C., p.83-102.

Dumbleton, M J (1962), "Investigations to assess the Potentialities of Lime for Soil Stabilisation in the UK", Road Research Technical Paper 64, HMSO, London.

Eades, J L and Grim, R E (1966), "A Quick Test to Determine Lime Requirements for Lime Stabilisation", Highway Research Record No. 3, TRB, Washington D.C., p.61-71.

Osula, D O A (1996), "A comparative Evaluation of Cement and Lime Modification of Laterite", Engineering Geology, Vol. 42, No 1, p.71-81.

Rogers, C D F, Glendinning, S and Roff, T E J (1996), "Lime Modification of Clay Soils for Construction Expediency", in press.

Sherwood, P (1993), "Soil Stabilisation with Cement and Lime: State-of-the-Art Review", HMSO, London.

Sweeney, D (1987), " Site Report on the Construction of a Lime Modified Subgrade on Highways No. 339-01 and 334-02", Department of Civil Engineering, University of Saskatchewan, Canada.

Thompson, M R and Eades, J L (1970), "Evaluation of Quick Test for Lime Stabilisation", Journal of Soil Mechanics and Foundation Division, ASCE, Vol. 96 SM2, p.795-800.

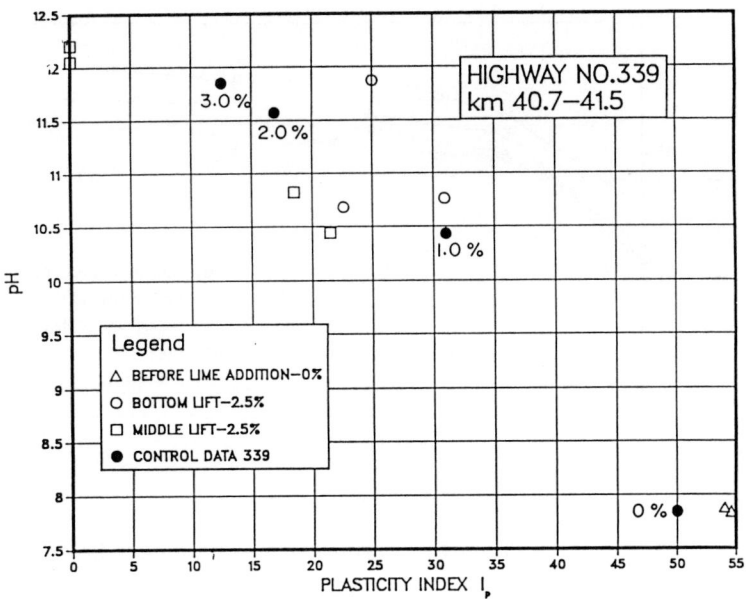

Figure 1. Graph of pH versus plasticity for subgrade treatment of Highway 339 (after Sweeney, 1987).

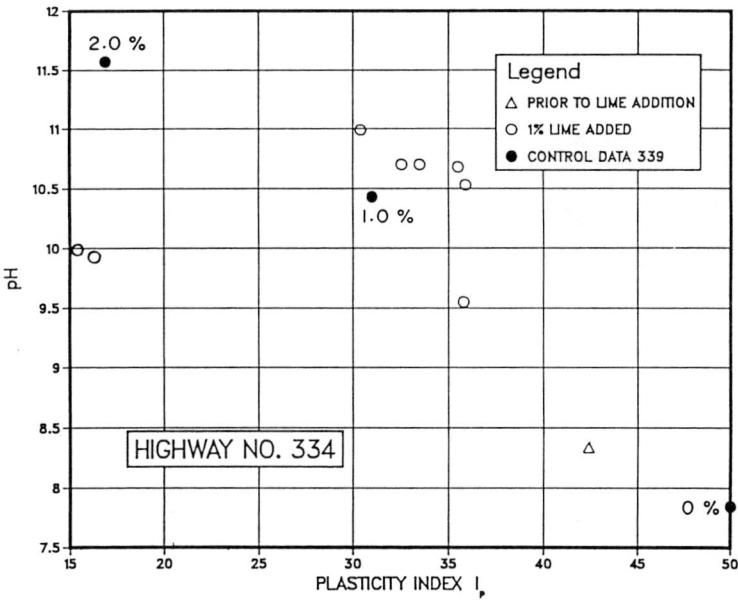

Figure 2. Graph of pH versus plasticity for subgrade treatment of Highway 334 (after Sweeney, 1987).

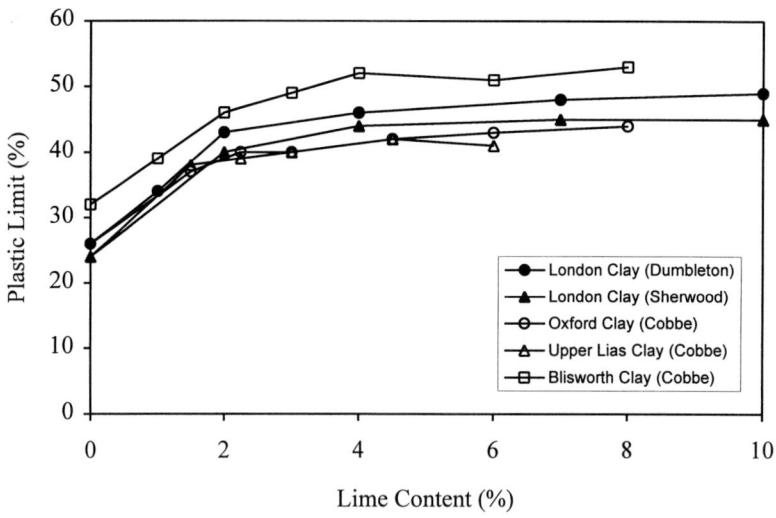

Figure 3. Change in Plastic Limit caused by lime addition to British Clays

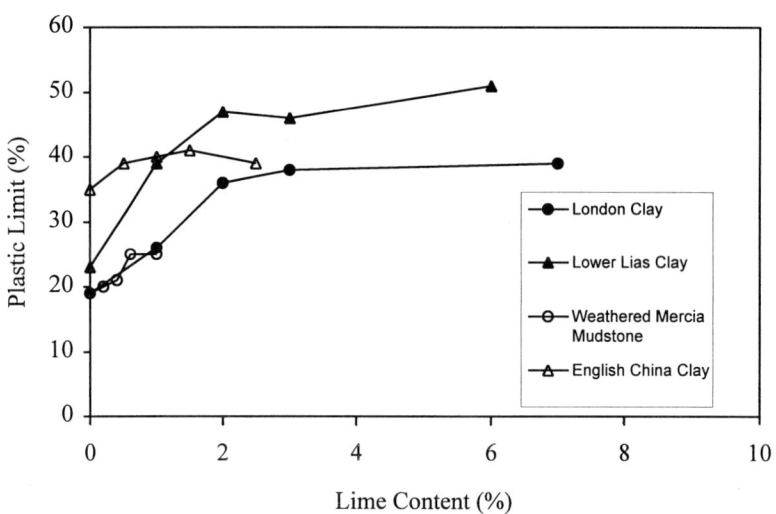

Figure 4. Change in Plastic Limit caused by lime addition to British Clays

114

Lime Treatment of Metal Contaminated Sludges

D.I.BOARDMAN
Department of Civil Engineering, Loughborough University
J.A.MACLEAN
Land and Water Services, Albury, Surrey, UK

INTRODUCTION

The Legacy of Industrial Wastes

Public awareness of the adverse effects of contamination on the environment has never been greater. Awareness has been developed by world environmental groups predominantly over the last decade. This has helped to apply pressure upon government bodies to act now with regard to pollution issues by introducing new legislation to control waste disposal. With the increasing demands for environmental responsibility, the European Community has set strict controls on the content and quantity of waste which can be disposed of at sea or applied to land. The economic consequence of this is greatly increased costs for industries producing toxic wastes and sludges as by-products of their production processes. In the United States of America the Government has a philosophy that means that the polluters must pay for the clean -up of contamination. In order for these industries to remain competitive with industries in countries where pollution control is not so rigorous, there is a need to develop economical treatment and disposal techniques. The scale of the problem is illustrated by the fact that the U.S. Environmental Protection Agency (EPA) estimates that, in 1981, the United States produced approximately 59 million metric tons (wet) of wastes that would be classed as hazardous (U.S. Environmental Protection Agency, 1980).

Sludge composition varies enormously from one industrial process to another. Typical sources of such wastes are copper, lead, and zinc smelting industries, duplicating and photographic equipment manufacturing industries, pharmaceutical industries and sewage treatment works. These are industries that are known to create these waste streams, which result in them being easily identifiable for potential pollution liability claims. Many other industries potentially contribute to the contamination of sediments in rivers, canals, lakes and reservoirs. The contaminant transport mechanism in these cases could be by water, air or by land but more likely a combination of all of these mechanisms. In this instance it can be difficult to ascertain who is the liable party. Hence when the removal of sediments from waterways is required, the costs of clean-up and ultimate disposal is often borne by the owner of the waterway.

Due to the variability of waste composition it is difficult to find a single treatment which fulfills all the requirements for safe long-term disposal. If wastes are to be disposed of, the toxic elements must in some way be contained within the waste disposal facility boundaries essentially forever, or at least losses kept so low that no harmful effects occur to the environment. Methods currently available and under research include land disposal, dewatering and landfill disposal, encapsulation and burial, and stabilisation and disposal on land or in landfill sites. There are intermediate processes which can be used to increase the physical, chemical and cost effectiveness of these processes. These methods include precipitating toxic metal species from the waste streams by altering the pH of the solutions. Toxic metals also tend to attenuate preferentially on small mineral particles. Thus it is possible to remove a large quantity of certain contaminants by removing specific particle sizes. This is often a pre-treatment when stabilising contaminated sediments. Stabilisation techniques generally involve mixing the waste with a chemical additive such as lime or cement. The combination of a high pH and the production of cementitious compounds create a solidified mass that can reduce the mobility of toxic elements in the waste.

Stabilisation methods tend to cost less than other methods, but application can be limited due the composition of the wastes and the limitations of the fixation processes. For example solids content is important as well as the form of the toxic species present in the waste. Some toxic species such as chromium have several valence states. Some states are less mobile in a stabilised waste matrix than others. If the treated waste is disposed of in an environment where water can permeate through the waste, certain species will become mobile. The presence of some contaminants can also impede the production of cementitious products which are associated with solidification and stabilisation methods.

The Purpose of this Paper

The use of lime is wide-spread in the United States (US) and patented proprietary solidification systems have been in use for some years (Tittlebaum et al, 1985; Malone and Jones, 1979). An example of such a proprietary system was assessed by the US Environmental Protection Agency. This particular patented system used bituminous flyash with lime added to produce a pozzolanic product. The amount of flyash added was related to the amount of total solids in the waste being treated. A final product with a high solids content of 80 percent was considered optimum and dewatering the sludge can often reduce the amount of flyash required (Landreth, 1982).

A new UK proprietary system shall be discussed with specific reference to the treatment of contaminated sediments, sewage sludge and its potential for adaptation to other industrial sludges and contaminated land.

This paper aims to highlight the fundamental parameters that affect lime stabilisation of metal contaminated sludges and sediments. An understanding of the parameters controlling the stabilising mechanisms will enable the assessment of new markets in the United Kingdom (UK) for waste stabilisation using lime. It will also discuss its practical application, concentrating upon UK experience.

SOLIDIFICATION AND STABILISATION PROCESSES

Many of the mechanisms associated with various solidification and stabilisation processes are well researched (Malone and Larson, 1983, Landreth, 1982, Weng and Huang, 1994, Betteker et al, 1986, Stanczyk, 1982, Dutré, 1995, Roy, 1993). General descriptions of several proprietary systems available in the US along with a guide to their application has been published by the US Environmental Protection Agency (Malone et al, 1980). It is known that certain inorganic and organic waste constituents in specific forms reduce the effectiveness of the processes. Malone and Larson (1983) state that it is often more appropriate to treat individual sludges produced by each individual stage of a production process. For example, in a metal finishing line, a chromium hydroxide waste from a plating operation was discharged along with a phenol waste from a surface cleaning operation. When attempts were made to solidify the sludge using a cement-based system, the chromium was retained, but the phenol levels were unacceptable. The sludge was treated with a strong oxidiser to destroy the phenol. This treatment, in turn, oxidised the chromium to produce hexavalent chromium. After oxidation treatment this new waste leached very high quantities of toxic hexavalent chromium. Due to the mixing of inorganic metal and organic waste streams, the solidification and stabilisation process became too complicated, resulting in the contaminated mixture being disposed of in a landfill site.

These problems can be overcome by thoroughly investigating waste composition and qualitatively assessing the likely mobility and hazardous properties of stabilised waste. This allows the design of an appropriate solidification and stabilisation admixture. The difficulties associated with traditional solidification and stabilisation binders such as lime or pozzolanic materials is that they are most effective at treating wastes containing inorganics. Binders used in the solidification and stabilisation of organic contaminants in wastes tend to be based on micro-encapsulation by thermoplastic materials, macro-encapsulation, organic polymerisation and solidification and stabilisation processes using organophilic clay. The use of organophilic clay can be limited by the fact that the clay chosen for the treatment is specific to the organic molecule being treated in the waste. Organic solidification and stabilisation is more expensive than inorganic solidification and stabilisation, and yields a smaller increase of the waste volume.

Both lime-pozzolanic agents and cement-pozzolanic agents have proven to be very cost effective. Waste materials found to be effective in solidification stabilisation applications are fly ash, cement kiln dust, waste lime tailings and clays from the manufacture of brick (Stanczyk *et al*, 1982).

The lime pozzolana solution

The supplies of lime in the form of quicklime (CaO) and hydrated lime (Ca[OH]$_2$) , and to a lesser extent cement, are relatively high around the world making lime an economically viable option for the treatment of wastes containing mobile inorganic contamination. When the use of lime was first assessed for treating sludges and sediments, benefits were perceived to be mainly physical. Due to the inherent nature of quicklime, it is a very effective dewatering agent. This is due to two mechanisms: the hydration reaction of CaO; and the exothermic nature of this hydration reaction:

$$CaO_{(s)} + H_2O_{(l)} \text{ Æ } Ca(OH)_{2(s)} + HEAT \qquad \text{Eqn 1}$$

This reaction produces approximately 17×10^9 Joules per kg of $CaO_{(s)}$. Equation 1 indicates that 1 litre of water is chemically used by the hydration of 3.117kg of quicklime. Assuming that there is an excess of water in the waste being treated, the 5.3×10^7kJ of energy produced will also evaporate water from the mixture. There is also the possibility that the energy produced by this reaction will promote endothermic reactions changing the nature of contaminated waste. If there are potentially volatile organic compounds present, they may be lost to the atmosphere due to the heat produced. There is also the possibility that the quicklime or the products of the quicklime hydration reaction will react with the contaminants in the waste directly. For example the production of ammonia gas when lime or any alkaline reagent reacts with Nitrogen compounds (e.g. fertilizers) can occur. Thus there may be additional benefits to the treatment as well as solidfication, however, the health and safety of the operators must be accounted for.

Pozzolans, which are sometimes used in conjunction with lime to improve the stabilisation of wastes, are defined as materials which are capable of reacting in the presence of water at ordinary temperatures, to produce cementitious compounds (Sherwood 1993). Natural pozzolans include volcanic ash and lava. Waste products from other industrial processes which are pozzolanic in nature include fly ash, pulverised burnt fire brick, burnt shale and ground blast furnace slag.

By far the most researched pozzolana used in stabilisation processes is fly ash. Weng and Huang (1994) investigated a specific fly ash with cement for treating industrial wastewaters. The chemical composition of the fly ash used is shown in Table 1.

The compositional data show a high percentage of silicates and aluminates. As with the addition of lime to a clay mineral, the high pH permits the formation of calcium silicates and aluminates It is this cementitious reaction which binds the particles together producing a solidified matrix. The high proportion of silica, alumina and iron oxide also implies that the fly ash will be a good metal adsorbent (Weng and Huang, 1994).

Table 1 Composition of an ASTM Class F Fly Ash (Weng and Huang, 1994)

Constituent	ASTM Class F Fly Ash (ASTM C618-89 specification)	
	Range (%)	Average (%)
SiO_2	36.83-50.54	43.57
Al_2O_3	25.07-38.57	31.44
Fe_2O_3	1.93-12.88	8.20
CaO	2.52-9.75	6.86
MgO	1.06-3.22	2.18
K_2O	2.41-5.52	4.45
TiO_2	0.80-4.76	3.28

The Effects Of Sludge Composition

Sludge composition is one of the fundamental parameters which can be used to assess the potential success of any stabilisation process. The main waste constituents and parameters for consideration are :

1. pH,
2. Solids content,
3. Metals content,
4. Sulphate, sulphite and chloride content,
5. Organic content,
6. Ammonia and
7. Cyanide.

The presence of some waste constituents such as sulphates, chlorides and organics can impede cementitious reactions. Wastes containing cyanide require oxidation and destruction of the cyanide prior to treatment. In some cases, ferrous sulphate can be utilised *in situ* to react with free cyanide. It is often the case that metals are the most abundant contaminants in wastes, hence their behaviour is very important. Firstly the oxidation state of inorganic elements or compounds affects the toxicity of the species and can also affect the mobility of the species due to the formation of different compounds of different solubility in different pH conditions. The pH of the treated waste is also very important as this can also dictate the ultimate disposal route. If a material is treated with lime and the pH is raised to pH 12.4, a landfill site will view this as a contaminated waste and there will be no real cost savings.

Dutré and Vandecasteele (1995) investigated the mobility of As (III) relative to As(V) before and after lime treatment of an industrial waste. The waste, originating from a metallurgical process in which copper is refined contained 42% by weight of arsenic. It was determined that nearly 95% of the untreated waste contained arsenic as As(III). Elemental arsenic is not toxic, but As (III) is 25 to 60 times as toxic as As(V) and several hundred times as toxic as methalated arsenic compounds. Various factors can affect the oxidation state of inorganic species. In this case, the waste also contained quantities of Fe and Fe(III). Fe(III) could oxidise As(III) to As(V), but in this case it was only present in small quantities so the effect was limited. The significance of this is that certain contaminants in a waste can be manipulated by the addition of another compound to reduce the potential toxicity of leachates produced. For example, not only can the oxidation state be adjusted, but various insoluble compounds can be created by addition of solutions containing Al, Ba or Ca ions (Wagemann, 1978) :

118

$AlAsO_4 : K_s = 1.6x10^{-16} = [Al^{3+}].[AsO_4^{3-}]$,

$Ba_3(AsO_4)_2 : K_s = 7.7x10^{-51} = [Ba^{2+}]^3.[AsO_4^{3-}]^2$,

$Ca_3(AsO_4)_2 : K_s = 6.8x10^{-19} = [Ca^{2+}]^3.[AsO_4^{3-}]^2$.

Where [ ] = Concentration of substance within brackets.
K_s = Solubility Product. The smaller the value the lower the solubility of the precipitate.

However, treatments such as these could limit disposal options. Due to the reversible nature of many of these reactions, the interaction of some landfill leachates could also transform the element into its most toxic and potentially into its most mobile form.

Griffin and Shimp (1978) carried out research investigating the effects of pH on cation and anion adsorption on clay liners in landfill sites. Two sources of anions and cations were utilised: pure reagent grade solutions of individual ions; and complex mixtures of ions from an actual landfill leachate. They found that in most cases more ions of a specific element were adsorbed from the pure solutions by the clays at a specific pH than from the leachate. This was primarily attributed to preferential cation or anion exchange and increased competition for adsorption sites by the leachate. At low pH the surface of the clays adsorbed more anionic species and at high pH the clays adsorbed more cationic species. The importance of the valence state is also emphasised. For example Cr (III) species were removed from solutions to a much greater extent than Cr(VI) species. The clay minerals removed 30 to 300 times more Cr(III) from solution than Cr(VI). There was also more extensive removal of As(V) than As(III). Hence these researchers indicate that for safer disposal of certain elements, it would be beneficial to convert them into the form that would be most strongly attenuated.

It was also noted by Griffin and Shimp (1978) that at various pH values different metal species are precipitated. Precipitation of hydroxides and/or carbonates of Cu and Pb from leachate occurred above pH 5, Zn above pH 7 and Cd above pH 6. Precipitation of Cr(III) as an amorphous hydrated hydroxide started to occur above pH 4.5, but no precipitate of Cr(VI) was detected between the pH range 1.0 to 9.0. Some As(V) precipitation was observed to occur above approximately pH 9. However a high pH did not guarantee removal of all metal species by attenuation and precipitation. The removal of Cu from leachate reached a maximum at about pH 7 and then decreased for values above pH 7. Griffin and Shimp (1978) attributed this to the amphoteric character of $Cu(OH)_2$ precipitates, which redissolve in basic solutions by forming $Cu(OH)_3^-$ anions. Thus it is clear that lime treatment will not be suitable for all metal species in all wastes. Each waste must be considered on an individual basis. Table 2 shows the general composition of several typical industrial wastes and indicates the large variability in constituent species and quantity.

Table 2 : The general composition of typical industrial waste streams and sludges (Landreth, 1982)

Sludge Description	Annual Production (Metric Tons, Wet)	% Solids	Density (kg/m³)	pH	Constituents >10,000 mg/kg (Dry)	Constituents 100-10,000 mg/kg (Dry)	Constituents 1-100 mg/kg (Dry)
Electroplating Sludge	50	32	1266	7.6	Ca, Cr, Cu, Fe, SO$_4$, Cl, Si	Be, Cd, Pb, Mg, Mn, Ni, Zn	As, Hg
Nickel-Cadmium Battery Sludge	100	40	1250	12.3	Ca, Ni, Cl, Si	Cd, Cr, Cu, Fe, Pb, Mg, Zn	As, Hg
Pigment Production Sludge	17,000	25	1170	8.4	Ca, Cr, Fe, Pb, Mg, SO$_4$,, Cl, Si	As, Cd, Cu, Mn, Ni, Zn, Hg	Hg
Chlorine Production Brine	3,000	59	1570	9.5	Ca, SO$_4$, Cl, Si	Cu, Fe, Mg, Mn, Ni, Zn, Hg	As, Cd, Cr, Pb
Glass Etching Sludge	2,000	47	1410	8.3	Ca, SO$_4$,, Cl, Si	Cu, Fe, Pb, Mg, Mn, Ni, Zn	As, Cd, Cr, Hg

CONSTRUCTION TECHNIQUES

It has been shown that lime treatment is a potential solution to the serious problem of sludge disposal. However, it does not offer a *carte blanche* solution for everything. Careful consideration of sludge composition is required before treatment options may be assessed. In practical terms sludge treatment is brought about by mixing combinations of additives such as lime cement, cement kiln dust and fly-ash with the waste sludge in quantities which are often dependent on the solids content of the sludge.

Betteker *et al* (1986) carried out extensive research into the solidification/stabilization of contaminated dredged material. Materials used for solidification / stabilization included :

1. Type I Portland cement.

2. A commercially available proprietary pozzolanic additive.

3. A proprietary polymer that is still in the research and testing stage and not currently available commercially.

The technologies were assessed for both physical and chemical stabilisation. Cement-based and pozzolanic-based technologies were used to convert contaminated sediment into a solidified product. Polymer was added to various processes to determine its effect on the leaching of selected organic compounds. Physical stabilisation was assessed by measuring the changes in unconfined compressive strength. Batch leaching tests were used to evaluate the chemical stabilisation. Leachates were analysed for polyaromatic hydrocarbons (PAHs), polychlorinated biphenyls (PCBs), arsenic, zinc cadmium, and lead. This research concluded that physical stabilisation of contaminated dredged material is a viable option, but no single process formulation proved to be effective in providing chemical stabilisation for all contaminants as a group. Portland cement proved to be the most effective at treating arsenic. More than half of the PCBs could not be detected in the leachate of Portland cement solidified sediment. Those PCBs that did leach were measured at less than one part per billion. Zinc, cadmium and lead were contained the most effectively by the proprietary pozzolanic additive, with no detectable concentrations of the metals in any of the leachates. The polymer additive was the most effective at containing the PAHs. Again more than half the PAHs were completely immobilised. Those PAHs that were not totally immobilised were found in concentrations four to five orders of magnitude lower in the leachates than in the untreated sediment. This indicates that some organic contaminants can be treatend by solidification processes and highlights the fact that every waste to be treated should be assessed independently to design the most appropriate stabilisation mix.

Contaminated sludges are the products of many processes and in most cases due to high water content of the sludge it is more economical to treat the sludge *in situ*. The mixing technique then becomes a problem. Stanczyk *et al* (1982) suggested the principle of a mobile mixing unit for solidification and stabilisation of waste streams and sludges. An outline design for the mobile unit is given in Figure 1.

Plate 1 shows a mobile solidification/stabilisation unit developed by Land and Water Services of Surrey in the United Kingdom.

The unit was initially developed for treating metal contaminated sediments from canals and lakes with quicklime. In 1994 Land and Water Services worked in conjunction with British Waterways treating the sediments of the Coventry Canal. The treatment rate of the wet sediment achieved during trials was 778 kg/min. The method proved to have great potential. It was concluded that to develop this technique in the most cost-effective manner the following actions were required :

1. The development of a standard set of blend ratios to suit site and material characteristics, to enable blend ratios and costings to be determined.

2. The establishment of new disposal options and alternative beneficial uses of the blended material to move away from conventional landfill options. Potential uses will depend on contamination, but could include tow path backfill, construction industry bulk fill, landfill capping, landscaping material and as an agricultural lime agent.
(Land and Water Services, 1994)

Recently the mobile treatment plant at Land and Water Services has been used to treat "uncontaminated" sewage sludge cakes as a post treatment to standard sewage treatment. The material has a solids content of approximately 27% but appears quite dry in nature. After the standard sewage treatment, the material contains a large quantity of flocculating agents which gives it its dry appearance and almost flexible properties. Due to the mixing method adopted for this particular process, the sludge cakes are effectively broken down into small pellets between 10mm and 30mm in diameter. These pellets are coated in a layer of fine cement kiln dust and fly ash. The reaction on the surface of the pellets effectively causes micro-encapsulation and solidification of the waste. Due to the nature of the treated waste, it can be sold to farmers to use on their fields as a lime agent and in general there has been a favourable response from the customer.

There are many other potential uses for a mobile unit. Figure 2 shows some of the waste waters and sludges which can be treated along with potential solidification reagents. The unit has the potential to be transported to industrial sites to treat stored sludges prior to disposal. The cost associated with this type of treatment depends on the type of the physical and chemical nature of the waste, reagents used for stabilisation, the quantity of reagent required for solidification, the ultimate disposal route for the treated waste, transportation costs and the quantity of waste to be treated. However, solidification and stabilisation techniques are used in the USA and are considered to be very economical.

CONCLUSIONS

It is clear that the problems associated with the treatment of contaminated sediments, waste streams and sludges are a growing concern. Combinations of lime and pozzolanic solidification and stabilisation techniques offer a potential solution which utilises waste products such as waste lime and fly ash. This is very cost effective but like any of the individual treatments discussed, it has its limitations. As a method for enhancing the physical properties of a saturated waste, lime addition is often very successful, allowing the treated waste to be transported easily and safely for disposal. Lime and pozzolanic treatment as a method for chemical stabilisation is far less certain. The process appears successful for many inorganic contaminants although not all. There is also some success with organic contaminants. The form of some contaminants can be manipulated by additives making them more suitable for chemical stabilisation, but it is clear that each waste must be considered and tested on an individual basis. Although the process may be successful in a controlled environment, the behaviour of the treated waste when finally deposited is uncertain. The chemical and biological interactions between a landfill leachate and a solidified/stabilised waste may affect the long-term success of the treatment. Research is required to assess the reactions of specific stabilisation and solidification additives with specific industrial wastes. With the interaction of industry and academia this research may lead to the creation of a UK database of solidification/stabilisation additives appropriate for the treatment of wastes from specific industries. This would help UK waste treatment contractors to assess the most efficient solidification and stabilisation treatment.

If lime and pozzolanic materials are deemed appropriate for the treatment of a specific waste, the benefits of mobile treatment plant is clear. Treatment *in situ* of waste can remove substantial quantities of water reducing the transport costs and the effects of transport on local communities. This is of benefit if the waste must go to landfill. However, material treated *in situ* can also be used locally depending on the type and level of contamination. For example a lime treated canal sediment could be used as a towpath backfill, and a treated sewage sludge could be used as an agricultural liming agent. For these methods to succeed, the mobile plant must be of flexible design, capable of working in awkward situations such as those on a canal, and be capable of handling wastes with varying physical properties. The mobile solidification and stabilisation plant developed by Land and Water Services of Surrey in the UK has proved this adaptability. The market for stabilisation/solidification processes and mobile plant is growing, but research is required to ensure the long-term stability and durability of the process in these new markets.

REFERENCES

Betteker, J M, Sherrard, J H and Ludwig, D D (1986), "Solidification/Stabilisation of Contaminated Dredged Material", Mid Atlantic Industrial Waste 18th Conference, Virginia, USA, p.253-273.

Dutré, V and Vandecasteele, C (1995), "Solidification/Stabilisation of Hazardous Arsenic Containing Waste from a Copper Refining Process", Journal of Hazardous Materials, Vol 40, p.55-68.

Griffin, R A and Shimp, N F (1978), "Attenuation of Pollutants in Municipal Landfill Leachate by Clay Minerals", Report Number EPA-600/2-78-157, U.S. Environmental Protection Agency, Cincinnati, Ohio 45268.

Land and Water Services (1994), "The Treatment of Saturated Dredgings using Quicklime", Report of Site Trials and Testing, Land and Water Services Ltd., Albury, Surrey, England.

Landreth, R E (1982), "Physical Properties and Leach Testing of Solidified/Stabilized Industrial Wastes", Report Number EPA-600/2-82-099, U.S. Environmental Protection Agency, Cincinnati, Ohio 45268.

Malone, P G and Jones, L W (1979), "Survey of Solidification/Stabilization Technology for Hazardous Industrial Wastes", Report Number EPA-60/2-76-182, U.S. Environmental Protection Agency, Cincinnati, Ohio.

Malone, P G, Jones, L W and Larson, P J (1980), " Guide to the Disposal of Chemically Stabilized and Solidified Waste", SW-872, U.S. Environmental Protection Agency, Cincinnati, Ohio.

Malone, P G and Larson, R J (1983), "Scientific Basis of Hazardous Waste Immobilization", Hazardous and Industrial Solid Waste Testing : Second Symposium, ASTM STP 805, R A Conway and W P Gulledge, Edited by the American Society for Testing and Materials, p.168-177.

Roy, A, Eaton, H C, Cartledge, F K and Tittlebaum, M E (1993), "Solidification/Stabilization of a Synthetic Electroplating Sludge in Cementitious binders containing NaOH", Journal of Hazardous Materials, Vol 34-35, p.53-71.

Sherwood, P (1993), "Soil Stabilisation with Cement and Lime", Transportation Research Laboratory, State-of-the-Art Review.

Stanczyk, T F, Senefelder, B C and Clarke, J H (1982), "Solidification/Stabilization Processes Appropriate to Hazardous Chemicals and Waste Spills", Hazardous Material Spills, Milwaukee, Wisconston, USA, p.79-84.

Tittlebaum, M E, Seals, R K, Cartledge, F K and Engels, S (1985), "State-of-the-Art on Stabilization of Hazardous Organic Liquid Wastes and Sludges", Critical Reviews in Environmental Control, Vol 15, No 2, p.179.

U.S. Environmental Protection Agency (1980), "Everybody's Problem: Hazardous Wastes", SW-826, Washington, D.C.

Wagemann, R (1978), "Some Theoretical Aspects of Stability and Solubility of Inorganic Arsenic in the Freshwater Environment", Water Research, Vol 12, p.139-145.

Weng, C H and Huang, C P (1994), "Treatment of Metal Industrial Wastewater by Fly Ash and Cement Fixation", Journal of Environmental Engineering, Vol 120, No 6, November/December, p.1470-1487.

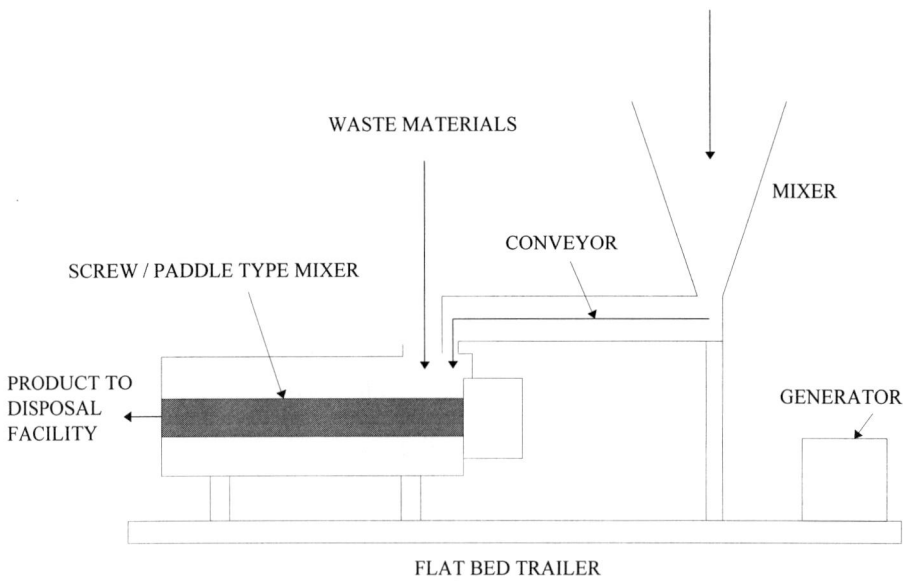

Figure 1 Proposed mobile solidification/stabilisation unit design (Stanczyk, 1982)

POUNDS OF REAGENT/
GALLON OF WATER

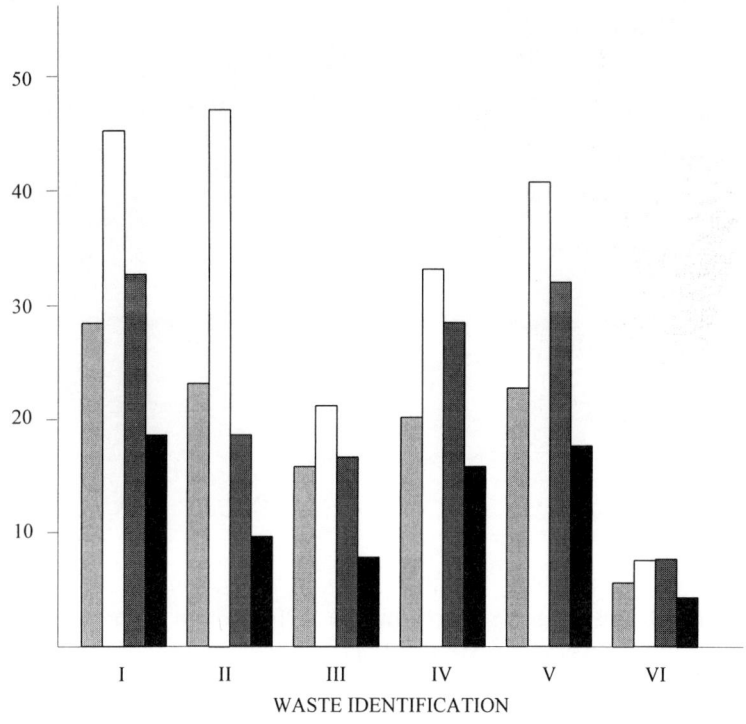

WASTE IDENTIFICATION

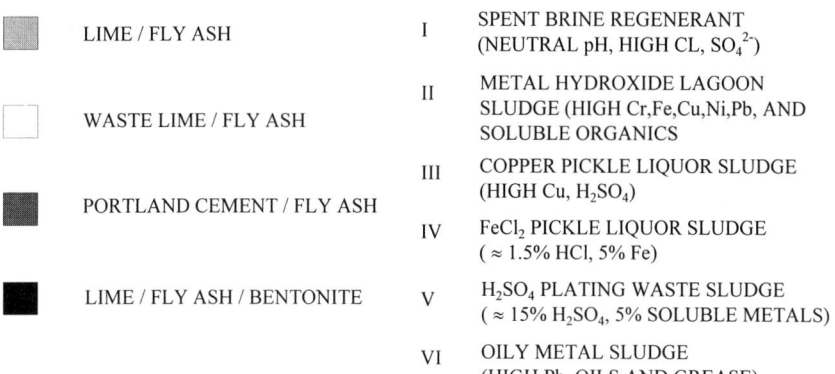

LIME / FLY ASH	I	SPENT BRINE REGENERANT (NEUTRAL pH, HIGH CL, SO_4^{2-})
WASTE LIME / FLY ASH	II	METAL HYDROXIDE LAGOON SLUDGE (HIGH Cr,Fe,Cu,Ni,Pb, AND SOLUBLE ORGANICS
PORTLAND CEMENT / FLY ASH	III	COPPER PICKLE LIQUOR SLUDGE (HIGH Cu, H_2SO_4)
	IV	$FeCl_2$ PICKLE LIQUOR SLUDGE ($\approx$ 1.5% HCl, 5% Fe)
LIME / FLY ASH / BENTONITE	V	H_2SO_4 PLATING WASTE SLUDGE ($\approx$ 15% H_2SO_4, 5% SOLUBLE METALS)
	VI	OILY METAL SLUDGE (HIGH Pb, OILS AND GREASE)

Figure 2 Reagent requirements for solidification of wastes for immediate transport (Stanczyk, 1982)

Plate 1   Land and Water Services mobile solidification plant

Deep Stabilisation Using Lime

S. GLENDINNING AND C.D.F. ROGERS
Department of Civil and Building Engineering, Loughborough University

INTRODUCTION

The use of lime as deep stabiliser has been pioneered in several overseas countries, notably Sweden, Japan and the USA, and can be divided into three main groups: lime columns, lime slurry pressure injection and lime piles. Lime columns are used primarily as an alternative to stone columns for stabilising soft soils, but have also been used to enhance slope stability. Lime piles are used for similar applications, whilst lime slurry pressure injection has its main application in slope stability. Choice of technique is dependent on stabilisation requirements, site restrictions and available equipment. This paper aims to outline the principles behind the improvement of soils using these techniques and discuss their potential application in the UK. Research has been carried out at Loughborough University into lime-clay mixing and, more particularly, lime piles for slope stabilisation. The paper thereafter aims to present the main findings of this research and, in so doing, to provide indicators of how the other deep stabilising techniques might be investigated and introduced into UK practice.

LIME COLUMNS

Principles

The term 'lime columns' refers to the creation of deep vertical columns of lime stabilised material. The apparent similarities between this technique and lime piles have resulted in considerable confusion in the literature. The general principle of the technique is to create, *in situ*, columns of intimately mixed lime and clay. The columns are constructed using a giant 'egg-beater' tool, shown in Figure 1. The tool drills itself into the soil while feeding lime into its centre to create a lime-soil mix containing typically 8-10% lime. Either hydrated lime (calcium hydroxide) or quicklime (calcium oxide) can be used, although the latter is generally preferred due to its dehydrating effect (Rogers and Glendinning, 1996). Reversal of rotation of the tool once the required depth is reached causes the mixed material to be compacted to create columns of typically 0.5-1.0m in diameter. The clay-lime reaction which takes place produces columns of material of greater strength and (initially) considerably increased permeability than the untreated soil. Thus, despite the fact that the strength improvement is derived from the material that occurs *in situ*, a two material system is established. As a consequence its applications are similar to those of stone columns.

In design, great emphasis is placed on the physical stabilising processes and the properties of clays required to promote effective stabilisation. Column shear strengths of as much as 5 to 20 times that of the parent material have been recorded, the increase in strength being time dependent. Permeability within the columns has also been raised by 100 to 1000 times immediately after mixing, although it is thought that the permeability will reduce in time as the stabilisation reactions progress and gel crystallisation occurs.

Applications

Broms and Boman have been the principal advocates of this method, publishing many papers on column construction, design and properties, and giving many examples of their successful usage. Broms and Boman (1975) give an introduction to the use and construction of lime columns, a complete study of design methods and applications being published by Broms in 1983. Applications were based on increased bearing capacities, shear strength and permeabilities. These include, essentially, foundations, with design methods being established by consideration of column strength and column-soil interaction.

Broms (1985) presents information on the use of lime columns in slope stabilisation. It was noted that the high induced pore water pressures produced by traditional piling techniques were reduced in the case of lime columns due to the quicklime's affinity for water. Design of pile installations was developed by considering the average shear strength (c_{av}) along a potential failure plane.

$$c_{av} = c_u (1 - a) + s_{col} \, a$$

where c_u is the initial shear strength of the soil (c_u rather than s_u is used as the author states that this value may be found by the shear vane test), s_{col} is the average shear strength of the stabilised clay within the columns and a is the relative column area, the ratio of the total area of the columns and the area of the stabilised soil.

Although pioneered in Scandinavia, lime columns have been used in several other countries including France where the technique has been adopted most notably for the stabilisation of railway embankments. The Japanese have also developed the 'deep mixing method' for use with lime (Terasai et al, 1979). Applications are reported to include slope stability, but no details are given.

The application of lime columns to British soils has been investigated by Corbet (1988), this being the only paper on the subject that considers the application for British use. Lime columns were assessed as a means of stabilising soft ground under embankments, the other forms of stabilisation being considered including drainage, vibro replacement, stone columns, geofabrics and conventional piling. Lime columns were seen as an alternative to stone columns but were judged unsuitable for the application required due to the inconsistency of strength gain and the lower than expected permeability found from laboratory trials. These results may be attributable, however, to sample preparation, sample storage and unsuitability of the tests, and the technique should not be discounted on this basis.

It is apparent that the columns have considerable potential for application in the UK and should be investigated rigorously in well designed experiments. Of crucial importance to their success are considerations of the intimacy and effectiveness of mixing, the water content of the clay to be treated in relation to the demand for water of the (relatively large added volume of) lime, and consequently whether hydrated lime, quicklime or possibly even lime slurry should be used. Such a decision will clearly depend on the application being considered and the initial condition of the soil.

LIME SLURRY PRESSURE INJECTION

Principles

Lime slurry pressure injection, as the name suggests, involves the introduction of a lime slurry into the ground under pressure (Figure 2). First developed in the 1960s, the technique forces the slurry into the pores and fissures in the clay, causing treatment by migration due to permeation of the slurry. It is claimed that treatment can take place without (significant) disturbance to the inherent clay structure, although hydrofracture during injection might be sought to enhance the slurry distribution. The lime forms a kind of matrix, enclosing areas of untreated clay. This is thought to prevent water movement between these areas, thereby creating wet-dry cycle resistance in swelling clays.

Applications

Applications include pavement remediation, building foundation construction and remediation, and embankment remediation, with typical injection depths of 1-2 m, 2.5-3.5 m and 3.5-15 m respectively. Injection typically takes place on a grid of 2.5 m centres using pressures of between 350 and 1400 kPa.

Blacklock and Wright (1986) review several US case studies using slurry injection and describe the development of field and laboratory tests to aid with the assessment of site suitability. In all cases quicklime was slaked on site. This was said to improve dispersion of the hydrate particles giving fine particle size and slower settling time, thus producing a higher surface area and hence reactivity.

Site evaluation has been achieved through specific test development, using adapted compression and shear strength tests for low strength soils and swell tests for expansive soils. Samples were dosed with 1% lime by dry mass, as this was thought to reflect field treatment. The sample preparation described was designed to simulate the ability of the treatment to repair and stop crack growth. Unfortunately the paper fails to explain adequately how the testing programme relates to field design or how the tests simulate site conditions.

The application of the technique in India has been reported by Bhattacharya and Bhattacharya (1989) where it has been used to treat soft and expansive soils under railway lines. The same principles as previously discussed were employed, with 'before and after' soil properties showing variable degrees of change.

One application for this technique is slope stabilisation. However, a review of the technique by Rogers and Bruce (1990) noted that the potential high pore pressures induced should be considered carefully when using the technique to this end.

There has been no record of the use of the technique in the UK, although cement slurry has been used for stabilising railway embankments and cuttings. It is unclear how the cement slurry produces significant improvement, although some lime migration from cement slurry is possible to a limited degree. The use of lime slurry pressure injection in the UK similarly warrants further investigation, although with the caveat discussed above and close attention to the means of construction.

LIME PILES

Principles

Lime piles are, very basically, holes in the ground filled with lime. Ingles and Metcalf (1972) show one method of lime pile construction, illustrated in Figure 3, in which a hollow tube is pushed into the soil to the required depth of pile and quicklime is forced into the tube under pressure as it is withdrawn. The pressure forces open the end of the tube allowing the lime to fill the cavity below. After each metre is filled, the end of the tube is closed and used to compact the lime forming the pile. The alternative method of construction is simply to auger holes to the required depth and subsequently, or via a central stem, to add quicklime and compact in layers. As an alternative to quicklime, hydrated lime, or more commonly lime slurry, is added to the augered holes, depending on the application and perceived stabilisation mechanism.

The literature concerning the process of stabilisation reveals two distinct applications for lime piles with two distinct sets of mechanisms. These shall be discussed separately herein.

Lime piles for soft soils

In China, Japan and Russia relatively large diameter (0.5-1.0m) quicklime piles have been used as a ground improvement technique for soft soil to improve its bearing capacity and reduce settlements of overlying structures (Wang, 1989; Kitsugi and Azakami, 1982; Tsytovich *et al*, 1971). Improvement is primarily attributed to pile expansion and clay dehydration (Wang, 1989; Kitsugi and Azakami, 1982; Chew *et al*, 1993). Water from the surrounding ground is consumed by the hydration of the (strongly hydrophilic) quicklime. As a consequence of the expansive nature of this reaction, lateral expansion of the piles is claimed to cause lateral consolidation of the surrounding clay. Arguments of stoichiometry (Rogers and Glendinning, 1994) suggest an alternative mechanism of clay consolidation (or densification) as a result of *dehydration* and *negative pore water pressures*, rather than generation of positive pore water pressures by outward movement of the pile. For this application a replacement method is used where and lime is forced into the soil under pressure. The current authors believe that this causes a significant proportion of the observed lateral movement.

Lime piles for slope stability

Lime piles have been used in the USA, Thailand and Austria as a method of slope stabilisation. The authors who propose their use for this application are relying upon the lime piles to stabilise failing slopes in stiff clays or weathered shales and mudstones and do not consider the 'expansion' mechanism. The research at Loughborough has shown this mechanism not to be significant when piles are being used in the context of slope stability. The lateral 'expansion', as evidenced by the increase in diameter of the pile, is reduced to a small percentage and in some cases zero. Due to the comparatively low water content of the soil in which the piles are constructed, the expansive nature of the hydration reaction is resisted and any tendency to expansion is manifested as compaction (i.e. is accommodated within the void space created by the granular nature of the lime). Piles in slopes are created by augering a hole and filling it with lime using a compactive effort that is small in comparison with the soil resistance, so that no forced lateral movement of the lime during construction is possible. Some authors do, however, mention the dehydration of the clay when using quicklime piles for slope stabilisation.

The reported stabilisation mechanism primarily concerns the migration of calcium ions from the pile into the surrounding clay and its subsequent stabilisation by lime-clay reaction. In order for this to occur both calcium ions and hydroxyl ions must migrate through the clay. In cases where lime slurry has been used to create the piles in the field, migration over significant distances is reported, probably as a result of hydraulic transport. 'Watering' of quicklime piles has also been suggested in order to improve hydraulic transport of ions. The measurements of ion migration are, however, suspect in many cases and in some cases migration is claimed without measurements to support the claim. Laboratory experiments at Loughborough and associated field studies have shown that the migration in clays is restricted to a relatively small distance (at least in the short term), as would be expected from consideration of the operation of clay barriers.

RESEARCH AT LOUGHBOROUGH UNIVERSITY

It is apparent from the literature that no definite set of stabilisation mechanisms has been established for lime piles. The main aim of the research at Loughborough was the identification and quantification of the different elements of the stabilising process when using lime for slope stabilisation. These elements will be discussed fully in the light of both laboratory and field results.

Laboratory studies

As a result of the apparently contradictory experimental evidence presented in the literature it was considered essential that each potential stabilisation mechanism be examined individually in appropriately designed laboratory tests. The various experimental procedures described by previous researchers, with appropriate adjustments in some cases, have been used to develop a series of laboratory investigations. An iterative approach was adopted whereby the results from one series of tests were used in the design of the next. This approach resulted in a range of tests including full-scale box tests in which lime piles were installed in clay samples, model scale lime pile tests and soil element tests.

The idea, described by several of the authors, of installing a small lime pile into a box filled with clay was used to study lime migration, pore water pressure changes and pile strength (Rogers and Glendinning, 1993). In addition to box model tests, the ideas of Noble and Anday (1967) were used to study migration and the strength changes occurring around a pile. 100mm diameter samples of clay were created by compaction at a known water content and central holes were created by pushing a 12mm rod into the clay. A smaller-scale version of this experiment was carried out specifically to investigate migration rates. Perspex tubes having an internal diameter of 30mm were filled with clays at different water contents. A 6mm diameter quicklime pile was placed at the centre of each sample and time taken for the ions to reach the edge of the tube (13mm) was recorded. This was achieved by integrating an acid-base indicator into the clay, which therefore changed colour as the alkaline hydroxyl ions passed through it.

As a result of the laboratory trials it was possible to identify, and to some degree quantify, the elements contributing to the stabilising mechanism. These elements are:

1. Generation of negative pore water pressure caused by the strongly hydrophilic quicklime in the piles drawing in water from the surrounding soil.

    For the purposes of design a nominal value of a reduction in pore water pressure of 20kPa is recommended for clays at a water content around their Plastic Limit. This figure is based on both laboratory and, importantly, field measurements. The duration of this pore water pressure reduction will be dependent upon depth, soil strata and ground water conditions. This mechanism is the most important in the short term but must be discounted in the longer term.

2. Overconsolidation of any existing shear zone(s) through which the piles pass, as a consequence of the negative pore water pressure, leading to increased strength of the clay in the shear zone (s).

    The results suggest that even modestly overconsolidated, pre-sheared clays show a significant improvement after overconsolidation by an effective stress increment of 20kPa. Results from testing English china clay indicated a nominal improvement in effective cohesion of whilst results from testing lower Lias Clay indicated an improvement of 4-5 kPa. Clearly, clay type, degree of overconsolidation and effective normal stress level influence the degree of improvement. The improvements may appear small, but when applied to shallow slips they can prove significant in terms of stability.

3. Dehydration of the clay surrounding the pile. Differential dehydration occurs in the clay mass, with the shear zone expected to be dehydrated to a greater extent.

    It has been shown that the reduction in water content in remoulded clay that is significantly wetter than Plastic Limit is radial in nature, i.e. reduces with distance from the pile, although prediction of the lateral extent of the reduction in the long term is not possible from the laboratory data since it depends upon several factors. Cracking has been shown to occur around piles both in the laboratory and in the field. Hydration of the quicklime via an external source of water would remove the benefit and this is possible at shallow depths via surface cracking and vegetation unless precautions are

131

taken. The potential for equilibration via an external source of water in the long term means that this mechanism should be largely disregarded.

4. Strength of the piles developed through hydration and crystallisation of the hydrated lime.

The pile strength recommended for design is 400 kPa. Such measured strengths were developed after a 90 day curing period in the laboratory, with a test carried out after 10 days yielding an undrained shear strength of 100 kPa, confirming that strength develops with time. Tests performed on piles placed in the field resulted in an undrained shear strength of 200 kPa on piles tested after 10 months and an undrained shear strength in excess of 400 kPa after a period of 20 months. Pile strength is the major contributor to the long-term stabilising mechanism, but should not be relied upon in the short term.

It should be noted that the expansion of lime in a closed system (i.e. below ground with no external source of water) solely compensates for the volume loss in the clay due to dehydration, as demonstrated by considerations of stoichiometry. Thus the volume of water taken in is almost precisely equal to the theoretical increase in volume of the pile. In practice the lime will tend to densify in the augered hole immediately on hydration, thus resulting in a compacted lime pile. This results in a net volume loss in the system as a whole and explains the presence of the ubiquitous tension cracks radiating out from the piles observed in both the laboratory and the field.

5. Improved clay strength around the piles produced by the lime-clay reaction.

The migration of calcium and hydroxyl ions and subsequent lime-clay reaction produces an annular zone of stabilised material restricted to approximately 30mm. The higher water content likely to be associated with a shear zone will promote lime migration and hence the most critical zone in terms of slope stability will be targeted by the treatment. In addition, it has been shown in the field trials that hydraulic transport via radial cracks causes the stabilised zone to increase significantly in size by creating a star-shaped pattern.

In order to determine a value of treated clay strength for design, mineralogical changes occurring in clays directly mixed with different quantities of quicklime were compared with those having quicklime migrated through them. The equivalent strength was found to be that achieved in an undrained triaxial test after mixing the clay with a lime addition equivalent to 2% above the Initial Consumption of Lime level (BSI, 1990) and curing for 28 days.

Field studies

In addition to the laboratory work, two field trials have been conducted. One of these employed two operators using a small 'Minute-man' drilling rig, which was used to create 2.5-3.0m long, 63mm diameter holes in heavily overconsolidated lower Lias Clay. The process, although slow, achieved steady progress and the 75 holes were completed within 8 working days. Lime was compacted using a 50mm diameter wooden stake after each 50 kg bag had been emptied into the hole and, on average, 2 bags were used in each hole. Spoil from the drilling process was used to create a platform for the next hole and to form the clay plug at the top of each pile. At the second trial site, which had far better access, a C200 drilling rig was used to create 150, 200mm diameter holes to a depth of 3-4m in London Clay. The same technique was used to fill the holes, the work being completed in 7 working days.

The successful completion of the first trial indicates the potential for use of relatively small scale plant, thus allowing the treatment of very inaccessible sites. This would be of particular benefit to densely vegetated sites where the use of larger equipment would prove difficult or would be prohibited due to environmental considerations. It may also prove useful in the treatment of cutting slopes adjacent to major communications links where lane closure would be very costly. The system therefore has potential as a low-cost means of treating traditionally 'problem' sites.

At the second trial site described above lime piles were placed into London Clay with a water content just below its plastic limit. Twenty months after installation 6 piles were exhumed to a depth of 1.5m and samples taken of the clay surrounding the piles. Testing for ion migration and changes in plasticity and water content is currently in progress. Observations made during the excavation and limited preliminary results shall be reported herein.

The clay immediately adjacent to the piles had become dry, hard and very strong. The width of the affected area was approximately 30mm. The clay surface that was in contact with the lime formed a very stiff dry crust, approximately 5mm in width. Beyond this the clay took on a whitish appearance and was still very strong and dry. The extent of this whitish zone varied, but was generally 20-25mm in width. Beyond this zone the clay became progressively softer and appeared to resemble that midway between the piles. Preliminary testing of the material suggested a significant reduction in plasticity index in this white zone, with some change up to 50mm away from the piles and little or no effect thereafter. This is in concurrence with the distance of migration and improved stiffness predicted from the laboratory experiments.

Radial cracking occurred around all sides of the piles excavated. The width of the cracks ranged from 2 to 5mm and the cracks were typically 50mm long, although in some cases were much longer. There were also very fine cracks between these larger cracks. In all cases lime appeared to have been forced along the cracks, possibly due to expansion of the lime on hydration but certainly aided by hydraulic transport. The clay surrounding the lime-filled cracks had formed a hard and dry crust approximately 15mm wide, which was considerably stronger than the surrounding soil. These observations mirrored those for the clay immediately adjacent to the piles.

It has been suggested by Handy (1994) that the radial cracks formed by the initial reaction fill with water, causing further expansion of the pile into the cracks and enhanced migration rates over a distance of up to 0.5 m. He has also experienced similar migration rates to those discussed above in the laboratory. Handy's results therefore appear to be in concurrence with the observations made herein.

The field trials demonstrated the suitability of the technique for use in shallow slope stabilisation. The technique proved itself particularly suitable for sites with limited access or environmentally sensitive sites due to its ability to utilise small-scale plant. The resulting data regarding pile strength, lime migration, radial cracking and lack of expansion largely confirmed laboratory results. It is, therefore , with increased confidence that future installations may be designed.

CONCLUSIONS

Lime columns have been used widely overseas for soft soil improvement and, to a lesser degree, slope stabilisation Although proved to be successful, the technique is yet to be adopted in the UK. Considerable investment in specialist plant would be necessary in order for this to happen and, more importantly, detailed and rigorous research onto its performance with heavily overconsolidated clays is necessary to prove its efficacy.

Lime slurry pressure injection has been pioneered in the USA with apparently successful results. Some doubts as to its suitability for slope stabilisation have been raised and no use in the UK has been established. Again, significant research is necessary to prove the benefits sought.

Lime piles have been used widely overseas for treatment of soft soils and slopes. Their use in the UK has been limited to slopes. Research at Loughborough has proved them to be an effective solution to the common problem of shallow slope instability. Stabilising mechanisms have been identified and quantified, allowing the design of full-scale trials. Recently, lime piles have been used commercially as part of stabilisation works on an embankment section of the London Underground railway network. This is discussed by Threadgold (1996).

ACKNOWLEDGEMENTS

The authors wish to thank the sponsors of the research into lime piles for their support, encouragement and technical input. The original research was jointly sponsored by the EPSRC LINK (Transportation, Infrastructure and Operations) and four industrial partners: Geotechnics Ltd., Cementation Piling and Foundations Ltd., Buxton Lime Industries and British Waterways. The advice from Howard Wyborn, the LINK co-ordinator is also gratefully acknowledged.

REFERENCES

Anon (1963), "Subgrade Improved With Drill-Lime Stabilisation", Rural and Urban Roads, October.

Ayyar, T S Ramanatha and Ramesan Koyilotan (1989), "A Study of Lime Columns in an Expansive Clay". Indian Geotechnical Conference, Visakhapatnam, Vol 1, p.185-189.

Bhattacharya, P and Bhattacharya, A (1989), "Stabilisation of Bad Banks of Railway Track by Lime Slurry Pressure Injection", Indian Geotechnical Conference, Visakhapatnam, Vol 1, p.315 -319.

Blacklock, J R and Wright, P J (1986), "Injection Stabilisation of Failed Highway Embankments", 65th Annual Meeting of the Transport Research Board, Washington D.C., USA.

British Standards Institution (1990), "Methods of Test for Stabilised Soils", BS1924, HMSO, London.

Broms, B B (1983), "Stabilisation of Soft Clay With Lime Columns", International Seminar on Construction Problems in Soft Soils, p.30.

Broms, B B (1985), "Stabilisation of Slopes and Deep Excavations with Lime and Cement Columns", 3rd International Geotechnical Seminar, Singapore, p.127-135.

Broms, B B and Boman, P (1975), "Lime Stabilised Columns", 5th Asian Regional Conference on Soil Mechanics and Foundation Engineering, Bangalore, p.227-234.

Broms, B B and Boman, P (1975), "Stabilisation of Soft Soil with Lime Columns", Ground Engineering, Vol 12, May, p.23-32.

Chummar, A V (1987), "Ground Improvement by Sand Lime Piles", Proceedings of the Ninth S E Asian Geotechnical Conference, Bangkok, Thailand.

Chew, H H, Takeda, T, Ichikawa, K and Hosoi, T (1993), "Chemico Lime Pile Soil Improvement Used for Soft Clay Ground", Eleventh Southeast Asian Geotechnical Conference, 4-6 May, Singapore, p.319-324.

Corbet, S P (1988), "Laboratory Trials for Lime Columns in Soft Clay", Lime Stabilisation '88 Symposium, London, British Aggregate Construction Materials Industries, March, p.64-79.

134

Davidson, L K, Demirel, T and Handy, R L (1965), "Soil Pulverisation and Lime Migration in Soil Lime Stabilisation", Highway Research Record No 92, Transportation Research Board, National Research Council, Washington D.C., USA.

Fohs, D G and Kinter, C B (1972), "Migration of Lime in Compacted Soil", Public Roads, Vol 37, No 1, p.1-8.

Handy, R L (1994), Personal communication.

Handy, R L and Williams, N W (1967), "Chemical Stabilisation of an Active Landslide", Civil Engineering, Vol 37, No 8, p.62-65.

Ingles, O G and Metcalfe, J B (1972), "Soil Stabilization Principles and Practice", Butterworths, Melbourne, Australia.

Katti, R K and Gupta, A K (1970), "Studies on the Diffusion of Lime in Expansive Soil", Proceedings of the Second S E Asian Conference on Soil Engineering, p.611-619.

Kitsugi, K and Azakami, R H (1982), "Lime-Column Techniques for the Improvement of Clay Ground", Symposium on Recent Developments in Ground Improvement Techniques, Bankok, p.105-115.

Lutenengger, A J and Dickson, J R (1984), "Experiences With Drilled Lime Stabilisation in the Mid-West USA", Proceedings of the Fourth International Symposium on Landslides.

National Lime Asociation (1985), "Lime Surry Pressure Injection", Bulletin 331, NLA Arlington, Virginia, USA.

Noble, D F and Anday, M C (1967), "Migration of Lime Deposited in Drill Holes", Virginia Highway Research Council Publication.

Rogers, C D F and Bruce, C J (1990), "Slope Stabilisation Using Lime", Lime Stabilisation '90 Symposium, Sutton Coldfield, British Aggregate Construction Materials Industries, 1 March.

Rogers, C D F and Glendinning, S (1993), "Stabilisation of Embankment Clay Fills Using Lime Piles", Proceedings of the International Conference on Engineered Fills, Newcastle-upon-Tyne, September, p.226-238, Thomas Telford, London.

Rogers, C D F and Glendinning, S (1994), "Deep Slope Stabilisation Using Lime", Transportation Research Record 1440, Transportation Research Board, National Research Council, Washington D. C., USA, p.63-70.

Rogers, C D F and Glendinning, S (1996), "Modification of Clay Soils Using Lime", Lime Stabilisation, Edited by C D F Rogers, S Glendinning and N Dixon, Thomas Telford, London.

Ruenkrairergsa, T and Pimsarn, T (1982), "Deep Hole Lime Stabilisation for Unstable Clay Shale Embankment", Proceedings of the Seventh S E Asia Geotechnics Conference, Hong Kong, 22-26 November, p.631-645.

Shanker, N, Babu, N and Maruti, G (1989), "Use of Lime Soil Piles for In-Situ Stabilisation of Black Cotton Soils", Indian Geotechnical Conference, Vol 1, p.149-153.

Terasai, M, Tanaka, H and Okamura, T (1979), "Engineering Properties of Lime Treated Marine Soils and Deep Mixing Method", Proceedings 6th Asian Regional Conference on SMFE, Vol. 1, p.191-194.

Threadgold, L (1996), "Slope Stabilisation using Reinforced Lime Piles", Lime Stabilisation, Edited by C D F Rogers, S Glendinning and N Dixon, Thomas Telford, London.

Tsytovich, N A, Abelev, M Yu and Takhirov, I G (1971), "Compacting Saturated Loess by Means of Lime Piles", 4th International Conference on Soil Mechanics and Foundation Engineering, Budapest, p.837-842.

Venkatanarayana P, Reddy P, Suryanarayana, Babu N and Ramesh V (1989), "Ground Improvement by Sand-Lime Columns", Indian Geotechnical Conference, Visakhapatnam, Vol 1.

Wang, W T (1989), "Experimentation of Improving Soft Clay With Lime Column", Int. Conf. on Engineering Problems of Regional Soils, p.477-480.

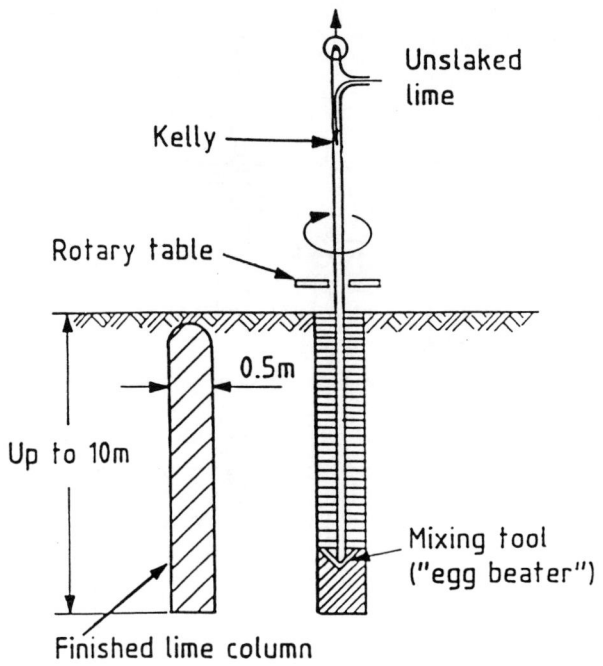

Figure 1   Procedure for construction of lime columns
(after Broms and Boman, 1979)

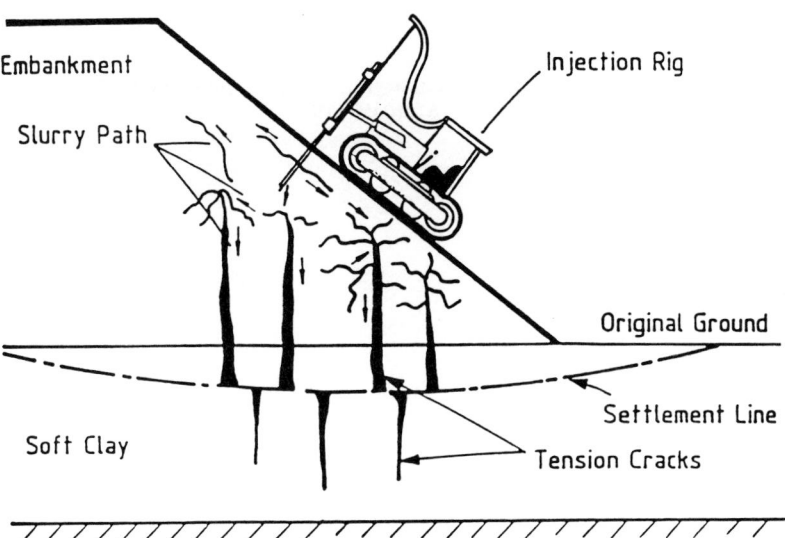

Figure 2   Slope stabilisation using lime slurry pressure injection
(after (US) National Lime Association, 1985)

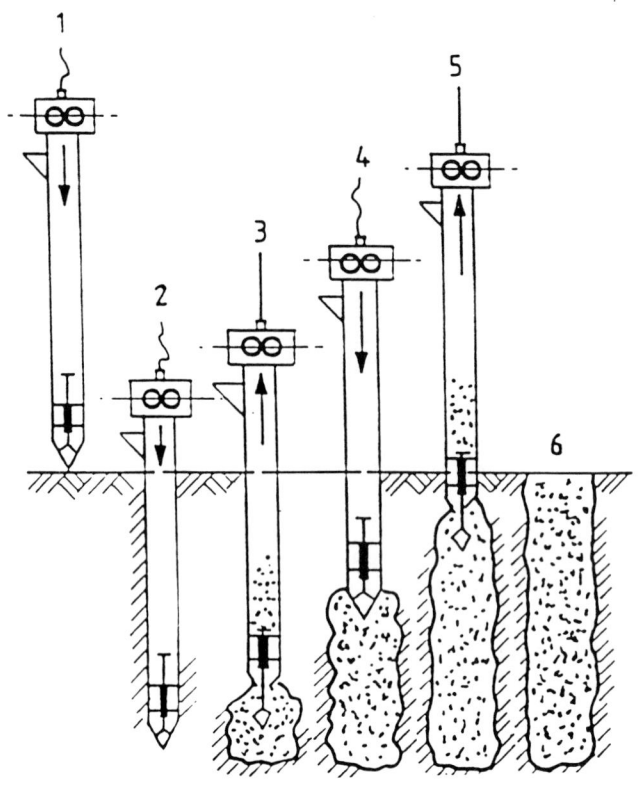

Figure 3   Procedure for construction of lime piles in soft soils
(after Ingles and Metcalf, 1972)

CASE STUDIES

Introduction

M40 - Lime Stabilisation Experiences

E A Snedker BSc (Hons), CEng, MICE.
Ove Arup and Partners.

Treatment of Silt using Lime and PFA to form Embankment Fill for the New A13

A Nettleton BSc (Hons), CEng, MICE.
Balfour Beatty Engineering Limited.

I Robertson BSc (Hons), PhD, MICE.
Hyder Consulting Limited.

J H Smith BSc (Hons), CEng, MICE.
Consultant to Powerbetter Developments Limited.

Slope Stabilisation using Reinforced Lime Piles

L Threadgold BEng (Hons), MEng, MICE, MIHT, MHKIE, FGS.
Managing Director and Chief Engineer, Geotechnics Limited.

Introduction

Previous chapters have drawn upon a combination of practical and research experience in order to present a 'state-of-the-art' review of lime stabilisation in the UK. The work presented is at the forefront of existing knowledge, with much of the work, especially from the previous chapter, yet to gain commercial credibility.

The aim of this chapter is to present three case studies from industry which point to current knowledge and experience used actively within the industry. As with everything, there is much to be learnt from 'having a go' and learning from the results. These papers aim to show, for three separate applications, this type of experience. It is hoped that the reader will gain valuable insight into and hence confidence to put in to practice, the principles of lime stabilisation.

The first of these papers deals with experiences encountered using lime stabilisation on sections of the M40. Unfortunately, if lime stabilisation is mentioned to most UK engineers, this 'famous' failure is the immediate response, and in some cases the technique is dismissed as far as consideration of its usage is concerned. However, as previously mentioned, it should be possible to learn from mistakes, given the ability to investigate the circumstances surrounding it fully. Indeed, it is usually only when a true failure occurs in engineering that a 'quantum leap' of understanding in that field is made. The aim of this paper is to present the facts about the M40 failure and the results of the subsequent investigation which point to clear reasons why it happened. It also points to measures to be made to prevent this happening again.

The paper describes how lime stabilisation was used as part of a multi-element capping layer on three of the four construction contracts, awarded between 1987 and 1989, for the 46km length of the M40 motorway between Banbury and Umberslade Junction. These contracts provided a sharp contrast between successful lime stabilisation and the deterioration of stabilised capping as evidenced by softening and swelling. Swelling of the lime stabilised capping by as much as 60% was noted during construction, lifting the pavement by a comparable amount. The pavement movements continued for some time at a diminishing rate. The areas of deterioration to the stabilised capping coincided with changes in the in-situ geological materials from which the stabilised capping was formed.

An analysis of the facts led to the conclusion that the deterioration was due to the formation of the mineral ettringite, the hydration of which is accompanied by large dimensional changes. A suitable environment for the formation of ettringite was found in the lime stabilised capping and its presence was confirmed by tests. After the discovery of the problem, the total sulphate levels were frequently found to be high. Comparison of the post-construction total sulphate values with the results available from the ground investigation showed considerable difference. The paper concentrates on the evaluation of the problem and highlights the necessary design decisions for future work.

The second paper presents practical experience of constructing embankments of lightweight fill for sections of the new £140m A13. The technique of lime modification (discussed by Rogers and Glendinning in Chapter 3) was used in conjunction with stabilisation of Thames silt from lagoons on the route of the road.

A total of 100,000m³ of silt was combined with quicklime and pulverised fuel ash (PFA) to produce lightweight fill. The silt, from the Port of London Authority dredging operation overlay soft alluvium and was classified as 'slightly contaminated'. Its usage in the fill mixture was estimated to save up to 40,000 truck movements on the already congested local roads.

This was achieved by preventing the need to remove the lagoons and import greater quantities of PFA.

The silt posed a serious problem for stabilisation being slightly contaminated, with 6% organics content, up to 100% water content and a high sulphate content. However, extensive laboratory and field testing was able to prove its engineering capabilities and its long-term chemical stability after treatment. The paper aims to report on the results of these tests and how the close co-operation of the contractor, consultant and specialist sub-contractor led to a highly successful project.

The final paper presents practical detail concerning the use of lime piles on London Underground's Jubilee Line. The paper describes the legacy of poorly constructed embankments experienced on sections of the Jubilee Line and the maintenance problems they pose. The paper describes how lime piles provided an extremely cost effective solution to stability problems on the section between Queensbury and Canons Park stations and how they are currently being applied to further sections.

Both the author of the paper and David Greenwood from the contractor, Cementation Piling and Foundations Limited, represented sponsors on the steering committee of the research on lime piles conducted at Loughborough University, the research background and overseas experience of which been reported by Glendinning and Rogers in Chapter 3. Early doubts about lime expansion led to the modified pile design described. Loughborough University is currently investigating the effectiveness of the first set of piles and results appear to be favourable.

This chapter will therefore 'de-mystify' the failure on the M40 and, it is hoped, finally allow this experience not to prevent its further usage in highway construction. It will also show how, given the courage to employ novel techniques, considerable cost savings are available without compromising design safety.

M40 - Lime Stabilisation Experiences

E.A.SNEDKER
Ove Arup & Partners, Coventry

THE SITE AND GEOLOGY

Lime stabilisation was used on three out of four contracts on the 46 km length of the M40 between Banbury and the Umberslade Interchange with the M42. In the south the route traverses both the Middle and Lower Lias geological series moving northwards onto the Keuper Marl series. Each of these series provided soil for lime stabilisation. Tracts of Alluvium and River Terrace Deposits occur throughout, but neither of these materials were involved in the lime stabilisation process. There are several areas of Glacial Drift deposits which occur along all four of the contracts, and the cohesive deposits were often used for stabilisation. These Glacial Deposits are from both the Western Drift and the Eastern Drift. The dividing point occurs towards the south end of the Gaydon contract, with the Eastern Drift Deposits to the south and Western Drift Deposits to the north.

DESIGN AND CONTRACT PREPARATION

The detailed design of the M40 Contracts between Banbury and Umberslade Interchange had been carried out on the basis that the Fifth Edition of the Department of Transport Specification for Road and Bridgeworks (1976) would be used for construction. That Specification had no direct inclusion for stabilisation work.

During the final preparation of the tender documents the Department of Transport published its Sixth Edition of the Specification for Highway Works (1986). As a consequence one contract (Warwick South) was invited based on the Fifth Edition Specification with no inclusion for stabilisation. A second contract (Warwick North) was invited again using the Fifth Edition Specification but with the addition of a draft version of the proposed future lime stabilisation specification. The two remaining contracts (Gaydon and Banbury IV) were invited based upon the Sixth Edition Specification which included stabilisation. The dates for invitation of tenders of the four contracts covered a one year period from February 1987 to February 1988. The demand for good quality rock in the Midlands was high at the time of tender and the three Contracts, which included the stabilisation option, were awarded including the use of a multi element capping incorporating lime stabilisation.

The ground investigations for this section of the M40 had been carried out at earlier stages in the design process (dating back to 1970). Consequently those investigations and the associated testing had not considered the possible use of lime stabilisation. It was apparent at the contract preparation stage that some areas might require replacement techniques as opposed to stabilisation *in situ*, where the material *in situ* was unlikely to meet various specification requirements. However, results from the investigation, as it stood, suggested that there was ample Class 7E material available on the site. Where particularly wet conditions were anticipated, or drainage continuity was required, a full depth granular capping was detailed.

Lime stabilisation. Thomas Telford, London, 1996

CONSTRUCTION CONTROL

Before stabilisation work commenced, compaction trials were carried out. The trials confirmed that the Contractors' proposed methods could meet the Specification requirements particularly in respect of pulverisation, rate of spread of lime, and the addition of correct amounts of water. They also demonstrated that the state of compaction could meet the levels aimed at in Table 6/4 of the Specification. Specimens were prepared from the trial areas to show that the required Bearing Ratio values could be achieved. Associated laboratory testing indicated the suitability of the constituents and determined Moisture Condition Value(MCV)/Moisture Content and Dry Density/Moisture Content relations for the mixed materials.

During the subsequent stabilisation process routine control tests were carried out on the properties of the constituents. Checks were made on the stabilisation methods and on the measurements to be determined during the process. Ultimately the necessary end product tests were carried out.

EVOLUTION OF THE PROBLEMS

Stabilisation work was successfully completed on the Warwick North Contract by the same Contractor who was constructing the Banbury IV Contract. No difficulty was experienced in the northern part of the Gaydon Contract. However, in November 1989, after some heavy rain and general lowering of temperature, the lime stabilised material at the southern end of this contract showed signs of deterioration. At this stage of construction the stabilised material was either exposed or covered by rock capping. The lime stabilised material was noted to have taken on a wavy appearance and had softened. Over the following months investigation showed extensive continuing deterioration and ultimately the soft material was removed and replaced by rock capping over large areas. The deterioration coincided with the use of glacial material from the Eastern Drift as the Class 7E material. To the north of this location Keuper Marl and Western Drift material had been used.

On the Banbury IV Contract, where construction was running ahead of the Gaydon Contract, the problem with the lime stabilised material manifested itself at a much later stage in the construction process. The Continuously Reinforced Concrete Roadbase (CRCR) had been constructed and some of the asphalt surfacing had been laid. The pavement and capping details for this contract are shown on Figure 1. Good progress had been achieved during excellent weather conditions in 1989 with the lime stabilisation being completed in August of that year. At the end of the winter period in April 1990 it was noticed that the cross sectional profile of the Southbound carriageway, over a section of the road which was in cutting, no longer conformed to the design. Two bumps in the longitudinal profile on the surface of the CRCR were also observed. Fine diagonal cracks were apparent in the CRCR. A level survey throughout the cutting showed that the carriageway had heaved in three distinct areas.

In the area of greatest heave, where a heave of 150mm had been noted on the CRCR surface, an inspection trench was excavated. Each layer of pavement was carefully removed down to the subformation and sampling and testing carried out. For comparison a second trench was excavated in an area which showed no sign of heave.

In the heaved area it became readily apparent that the lime stabilised material had expanded appreciably and was very soft and wet. Locally CBR values of less than 1% were measured and moisture contents of 55-60% were recorded giving a consistency index of just over 0.5.

The other layers of the pavement, the CRCR, sub-base and rock capping material showed no sign of any problems and the underlying sub-formation was sound. The depth of the lime stabilised material had increased by an amount which directly reflected the apparent heave determined from the level survey.

It became clear that the cause of the heave to the carriageway was directly attributable to the expansion of the lime stabilised capping material. The expansion had reached 60% of the original thickness. Level surveys were subsequently carried out for the entire length of the contract. The results showed that heave had occurred to a varying degree at many other locations. The heaved areas only occurred above the lime stabilised material and not above the full depth rock capping. At that stage most of the heave appeared to occur close to the interfaces of full depth rock capping and the multi element capping layer. The problem on the Banbury IV Contract had not occurred where Middle Lias had been used as the Class 7E material. It appeared to be generally, but not exclusively, confined to areas of cutting formed through Lower Lias material.

INVESTIGATORY WORK

As the defective lime stabilised material on the Gaydon Contract was removed the Banbury IV Contract offered the opportunity for more detailed investigation of the problem.

Following the discovery of heave on the Banbury IV Contract a programme of sampling and testing was formulated to determine: the cause of the expansion of the lime stabilised material; the extent of the problem and risk to the pavement throughout the Contract; any potential for further expansion of the lime stabilised material; and to allow the Contractor to continue surfacing operations without risk.

Samples of both the lime stabilised material and the parent sub-formation material were taken from all lengths of lime stabilisation, both in areas where there was evidence of heave and in areas showing little or no heave.

A testing programme was initiated, which included attempts to simulate the problem in the laboratory. A testing regime was imposed on the in-situ lime stabilised material and sub-formation material, together with the groundwater. The testing included classification tests, chemical tests, moisture content and density determinations, swelling tests, bearing ratio tests, X-ray diffractometry(XRD), electron microprobe microanalysis, and scanning electron microscope photomicrographs. Samples for testing were taken from locations throughout the contract where lime stabilisation had been used.

Sampling of the material was achieved by excavating a pit in the verge or the central reserve at the chosen chainage. The lime stabilised layer and the sub-formation were then excavated horizontally beneath the CRCR for a distance of some 750mm and the face of lime stabilised material and sub-formation material thus exposed was sampled. The thickness of the lime stabilised layer was established as accurately as conditions permitted. Two undisturbed samples were taken from the lime stabilised material by the process of jacking in a CBR mould horizontally. Frequently the sub-formation below the stabilised material could not be sampled in the same way due to its stiffness and so a bulk sample was taken following a measurement *in situ* of the CBR using a field assessment cone penetrometer. The samples obtained were subjected to laboratory testing to obtain the CBR of the undisturbed sample, CBR and expansion of the soaked undisturbed sample, moisture content, dry density, pH value, total sulphate content, soluble sulphate content, organic matter content and Atterberg limits.

In addition some of the samples from the field were tested for any lack of hydration of the lime. The material was mixed with water in a thermos flask fitted with a thermocouple to detect any heat of hydration. No change in temperature was detected.

Analysis of the considerable amount of data produced along the Contract indicated that:

1. The total sulphate contents of the unstabilised sub-formation material were generally higher than the values recorded for similar material in the ground investigations (see Figure 2);

2. There was a strong relationship between the moisture content (and related swell) and the total sulphate content in the lime stabilised material (see Figure 3);

3. The total sulphate values in both the lime stabilised material and its parent sub formation material were generally similar;

4. Considerable increases in moisture content, since compaction, had accompanied heave in the lime stabilised material (see Figure 4). This resulted in reductions in the Bearing Ratio and Dry Density of the stabilised material.

5. In the area of heave examined, the expansive mineral ettringite was identified in the stabilised clay whilst the unstabilised Lower Lias Clay was found to have the constituents to form ettringite and thaumasite (see Table 1(c)).

6. In the areas which were examined where no heave was evident, only limited, small amounts of ettringite were identifiable in the stabilised clay. The unstabilised clay did not have the constituents readily available to form ettringite (see Tables 1(a) and (b)).

The scanning electron microscope photomicrographs taken of samples from Banbury Contract of both stabilised and unstabilised Lower Lias Clay, showed the formation of ettringite crystals in the lime stabilised samples, one of which had a measured total sulphate content as low as 0.37% (see Plate 1). This agrees with the research undertaken by Mitchell and Dermatas (1990) which has shown that the presence of as little as 0.3% of sulphate can result in the formation of ettringite.

Previous problems in the UK involving deterioration of lime stabilised soil had generally been attributed to inadequate compaction. However, having carried out a compaction trial on each contract and employed regular supervision on the methods used, it was difficult to suspect the compaction method.

On the Banbury IV Contract the stabilisation had taken place during a very warm summer. It had been hypothesised that, perhaps under the unduly dry conditions, there had been insufficient water available to slake the quicklime. Unslaked lime therefore may have been incorporated into the capping layer and with a subsequent inundation of water the heat energy generated may have been sufficient to be converted into a lifting mechanism. This hypothesis was subsequently rejected. Calculations showed that under ideal conditions hydration of the quicklime was likely to generate a temperature rise of little more than 6°C. Confirmatory measurements of the temperature generated were made in the laboratory. Soil samples were prepared with a high moisture content deficit, mixed with quicklime, compacted and soaked to simulate a very extreme case of delayed hydration. The highest temperature increase recorded was 4°C above room temperature and within 24 hours the temperature had fallen back to room temperature. In addition, the quantity of water required to hydrate the quicklime is a small percentage of the moisture content of the soil. It was difficult to imagine that most of the hydration process had not taken place during the initial mixing of the lime with the soil *in situ*

Table 1 X-ray diffraction analysis

1(a) Middle Lias — Area of No Heave

MINERAL	PARENT SUB FORMATION			LIME STABILISED	
Quartz	4.4	9.6	4.0	10.3	13.9
Feldspar	2.2	6.9	1.5	2.4	2.9
Kandite	43.6	20.6	48.3	39.4	38.5
Mica	22.2	14.5	27.8	19.4	15.0
Illite	18.2	29.0	10.8	18.6	19.9
Smectite/Chlorite	9.0	19.1	4.9	7.6	7.1
Calcite	0.0	0.0	0.0	0.5	2.3
Dolomite	0.0	0.0	0.0	1.3	0.0
Gypsum	0.0	0.0	0.0	0.1	0.0
Pyrite	0.0	0.0	2.4	0.0	0.0
Ettringite	0.0	0.0	0.0	0.0	0.0

1(b) Lower Lias — Area of No Heave

MINERAL	PARENT SUB FORMATION				LIME STABILISED		
Quartz	5.0	6.2	6.7	6.0	6.4	5.6	5.9
Feldspar	0.1	0.0	0.6	0.5	0.5	0.4	0.2
Kandite	31.6	30.1	29.2	26.0	29.7	24.5	31.4
Mica	6.2	7.8	5.7	4.9	4.2	4.6	7.2
Illite	25.1	20.6	21.6	27.0	23.8	21.8	21.9
Smectite/Chlorite	10.9	9.3	8.1	7.0	7.3	5.8	7.2
Calcite	12.3	14.0	14.0	13.3	13.2	22.5	14.5
Dolomite	3.8	5.1	5.9	6.0	6.0	5.6	5.2
Gypsum	0.0	0.0	0.0	0.0	0.0	0.0	0.0
Pyrite	4.7	6.4	7.7	7.1	5.4	6.6	4.7
Ettringite	0.0	0.0	0.0	1.6	3.0	2.2	1.3

1(c) Lower Lias — Area of Significant Heave

MINERAL	PARENT SUB FORMATION					LIME STABILISED		
Quartz	10.6	10.2	11.4	8.5	10.0	10.1	8.3	6.3
Feldspar	0.5	0.0	0.0	0.2	0.3	0.5	0.5	0.4
Kandite	15.2	19.9	14.9	15.1	13.7	18.7	17.9	12.5
Mica	0.0	0.0	0.0	4.9	4.3	0.0	0.0	8.3
Illite	35.1	26.5	29.7	22.6	18.3	25.6	26.6	24.2
Smectite/Chlorite	9.5	13.0	11.7	11.0	10.3	4.5	6.1	3.4
Calcite	19.3	20.0	21.9	27.4	30.4	15.9	17.0	20.8
Dolomite	7.3	7.2	7.3	5.9	8.6	6.8	5.1	3.1
Gypsum	2.2	2.8	2.7	3.9	3.5	1.1	0.4	0.2
Pyrite	0.0	0.0	0.0	0.0	0.0	0.0	0.0	0.0
Ettringite	0.0	0.0	0.0	0.0	0.0	16.5	17.6	20.2

(Approximate percentages)

and the subsequent watering and re-mixing prior to compaction. The regular site control testing did not suggest that the mixed material had been compacted dry.

Because of the early hypothesis that there had been lack of hydration of the lime, the initial groups of simulation tests were aimed at reproducing a swell mechanism in the laboratory based upon possible dry compaction and lack of hydration of the quicklime. Although some swell was recorded the tests did not prove that lack of hydration was causing large swells in the material. Any heat generation due to hydration occurred quickly upon mixing the soil with lime and soon dissipated. Although a swell was recorded this generally proved to be no more dramatic than that produced from similarly compacted unstabilised specimens. For material originating from areas of low sulphate soil any swell was generally complete with 10 days. For material originating from areas of high sulphate soil the swell progressed at a much slower rate. This appears to agree with a similar observation made by Hunter (1988). There was also no identifiable trend to show that the use of quicklime, as opposed to hydrated lime, gave any consistently higher swell.

At a later stage of the investigation into the cause of swelling a further set of laboratory tests were carried out in an attempt to simulate the formation of ettringite and subsequent heave. The specimens prepared for this purpose showed a continuing swell over a long period of time (25 months) although the magnitude of the final swell did not approach the maximum proportions observed on site. The following points were noted from these tests:

1. The specimens did not saturate throughout as occurred in the field, and the observed field consistency was not reproduced. More water was absorbed by the top of the specimen than the bottom. Even so, two of the specimens swelled by over 20% in that time.

2. The presence of ettringite/thaumasite was observed in the specimens through X-ray diffraction tests, as was gypsum. This suggested that further formation of ettringite/thaumasite may have been possible given more conducive conditions.

3. When the temperature of the curing tank water fell below 15°C, due to a power failure, each specimen showed an apparent small but sudden increase in swell.

4. As expected from the reaction model set out by Hunter (1988), the rate of increase of swell declined as the pH value declined.

It was evident at the end of the tests that the conditions which occurred in the field had not been reproduced in the swell tests. It was considered that neither the water access to the specimen nor the field temperature changes had been reproduced in the laboratory.

OTHER EXPERIENCES AND RESEARCH

Lime stabilised pavement foundations have been used successfully for a long time. However, in the UK breakdown of lime stabilised material accompanied by excessive moisture content increase was reported by Caerns and Noakes (1988). It is understood that a similar case occurred in Cambridgeshire at about the same time. Heave combined with an increase in moisture content was also recorded on the M25 in 1983. These cases appear to have been attributed to insufficient compaction.

It is only during relatively recent use that cases of substantial heave of lime stabilised materials have been publicly reported and these were from the USA (Mitchell, 1986; Hunter, 1988). The reported magnitude of heave of lime stabilised materials had been substantial, occurring some time after construction, when water had penetrated the stabilised layer. The common features of all the reported cases of heave in the USA were the presence of sulphates, in either the soil or the ground water, and that the process of heave had continued over a period of years. Subsequent research at the University of California at Berkeley has examined the problem of the heave of lime stabilised soils in some detail (Mitchell and Dermatas, 1990).

The detailed chemistry of the reactions between the lime, the sulphates and the soil minerals is very complex and depends greatly on the pH of the soil, the temperature and the water conditions. What has been established is that, where substantial swelling of lime stabilised material takes place, it is associated with the formation of the mineral ettringite, one of the calcium-aluminium-sulphate hydrates. The hydration of ettringite is accompanied by large dimensional changes as the water of crystallisation is incorporated into the mineral structure. Ettringite often occurs as long needle like crystals and, where the interparticle clay bonding is weak and the ambient stresses are low, these crystals can push apart the clay particles. In addition, the crystal growth will probably weaken the physico-chemical bonding between clay particles in the soil and permit them to swell under the low ambient stresses and in the presence of water. The referenced papers suggest that once ettringite has started to form it continues to grow, provided that the necessary conditions and constituents remain available, until the temperature of the system falls below 15°C. Then, provided that the pH value remains high (above 10.5), the dissolution of any available carbonates together with available silica allows a substitution of silica for alumina and carbonate for sulphate leading to a conversion from ettringite to thaumasite, a mineral with similar expansive properties to ettringite. Unfortunately, there is some difficulty in distinguishing between the two minerals in X-ray diffraction work. Both minerals cannot form without an abundance of water. The conversion to thaumasite can produce a quantity of secondary gypsum. The reaction model indicating the reaction leading to the formation of ettringite and thaumasite is described in detail by Hunter (1988).

The observed behaviour of the lime stabilised soils on the Banbury IV Contract was consistent with that reported by Hunter (1988) and Mitchell(1986).

ASSESSMENT OF THE INVESTIGATORY FINDINGS

There was a very marked difference between the total sulphate values recorded in the ground investigation and those results obtained from below the pavement after it had been constructed. The difference can be seen in the histograms in Figure 2 which compare results from the Lower Lias material directly beneath the lime stabilised layer with the Lower Lias material tested in the ground investigation. A similar increase above the ground investigation values was also evident in the lime stabilised capping.

The discrepancy between the two sets of results based on a large number of samples raised the following considerations:

1. Could the ground investigation data have been incorrect?

2. Could the post construction tests have been incorrect?

3. Had there been an influx of sulphates into the material between the ground investigation stage and post construction?

The preconstruction total sulphate content results were obtained from five separate ground investigation contracts carried out by three different ground investigation contractors over a period of 18 years. These investigations (along various alignments in the route corridor) accumulated 145 total sulphate content results in the materials on the M40 Banbury IV section of which 81 tests were obtained on the as built alignment.

The ground investigations were carried out to the Department of Transport specification for ground investigation work current at the time. It is difficult to see how five investigations could all have provided an incorrect set of results. The post construction sulphate results were all carried out at the central laboratory of a reputable materials testing consultant. There was no evidence to suggest that these results were incorrect.

There was a considered possibility that additional sulphates could have arisen by transfer from external sources either from groundwater or from adjacent material, perhaps transmitted within the water percolating through the constructed pavement or used in the stabilisation process. Some work was set in motion to investigate this possibility. This included testing of groundwater obtained from the initial heaved area, testing of the rock capping material and testing of water flowing through appropriate culverts. Discussions took place with farmers adjacent to the motorway regarding their use of fertilizers and consideration was given to the specified application of fertilizers to the side slopes and verges. It was concluded from the information obtained that the additional sulphates could not have resulted from these sources.

The higher post-construction total sulphate levels were evident both in the lime stabilised material and the parent sub-formation material directly below. In the course of expansion, the affected lime stabilised material had taken up large quantities of water and its resulting moisture content was generally higher than the sub-formation material below. The sub-formation moisture contents remained broadly within the range prevailing at the time of construction (see Figure 4). An analysis of total sulphate content and soluble sulphate content of both stabilised and sub-formation materials was made. It was apparent that the soluble sulphate content remained at a constant level and was unaffected by either total sulphate content or moisture content. It was therefore concluded that the increases in sulphate level were due to the presence of additional insoluble sulphate. On the basis of other testing this was most likely to be gypsum, which had been absorbed into the lime stabilised layer. However, the high total sulphate content was present in both the lime stabilised capping and its parent material. It was likely, therefore, that any high total sulphate content was in existence at the time of stabilisation.

The possibility of an increase in the level of sulphates in a soil under certain conditions has been recorded by a number of authors. Hawkins and Pinches (1987) have furthered research into this aspect of behaviour and their results were considered to be relevant. The general reaction causing this increase involves the oxidation of pyrite which releases acidic fluids which then take available calcite into solution. A further reaction of calcium with the resulting sulphates produces calcium sulphate which, when hydrated, forms gypsum. The detailed reaction involved is set out by Hawkins and Pinches (1987).

Hawkins and Pinches (1987) also make the point that although the BS1377:1975 test for total sulphate is repeatable it only measures the value at the time of test. No account is taken of the potential for sulphate production which can occur later if air and water are able to oxidise soils containing sulphides. What was more important, the authors attempted to attribute the time scale involved in such a reaction and to quantify the sulphate increase which can occur. Their tests indicated a three to four fold increase in total sulphate value over 74 days which occurred in a linear manner. They have also suggested, from their test results, that the rate of increase is greater as the temperature increases. This work provides more detail and substantiation of the statement by Gallios and Horton (1981) that the formation of sulphates can be artificially induced in a matter of weeks simply by wrapping appropriate cores in plastic sheeting, or mylar, and leaving them in a warm place. The oxidisation process which can occur with depth, given the availability of oxidising agencies, is indicated by Russell and Parker (1979). They show profiles of relevant mineral content with depth for Oxford Clay to demonstrate the effect.

It was considered that such a mechanism occurring in the Lower Lias provided the explanation for the disparity in total sulphate results. The excavation for the earthworks would have provided the oxidising agencies to material which had not previously been subjected to them. In addition most of the excavated material was subjected to high summer temperatures.

Unfortunately there was no requirement to identify sulphides in the soil during the design process. The Lower Lias sediments are mostly marine shales and mudstones deposited under relatively shallow shelf conditions. It was noted by Shaw (1981) that during early burial biogenesis under reducing conditions iron pyrites may form in marine muds by alteration of iron oxides and oxyhydroxides. The sulphides can form in a reducing environment within the sediments. A further comment by Gallios and Horton (1981) states that pyrite is ubiquitous in marine mudstones. In addition, published mineralogy results (Perrin, 1971) from the blue grey and brown grey clays of the Lower Lias showed the presence of pyrites. The overall suggestion was that disseminated pyrite could be expected throughout the Lower Lias as well as the concentrated band of pyrite often noted immediately above the White Lias.

An examination of the XRD results obtained in the post construction testing work did identify the presence of some pyrite (see Tables 1(a),(b) and (c)).

Firstly in the Middle Lias materials, an area unaffected by heave, pyrite was only evident in one XRD sample. However in all of those samples calcite and dolomite which would be required in any oxidation process were virtually absent, as was the presence of gypsum.

The Lower Lias material results were more revealing. Samples had been taken in an area where heave had occurred and also where no heave had occurred. In the area of no heave, pyrite of up to 7.7% was identified. Calcite and dolomite were both present but no gypsum was recorded. It was reasonably suggested that in this case the environment for oxidation had not been present. In the area where heave had occurred calcite and dolomite were present but no trace of pyrite was found. However, up to 2.8% of gypsum was noted. In this case there had obviously been the opportunity for pyrite to oxidise and form gypsum.

If oxidation of sulphates had taken place in geological time it would be expected that the oxidation profile as indicated by total sulphate content with depth below pre-construction original ground level would be very similar both at ground investigation stage and at post construction stage. A similar number of results were available at both stages and these are shown in Figure 5.

It is apparent that although there may be a peak of total sulphate values created by a few samples around 3 metres below original ground level there is a very significant increase in the actual total sulphate values recorded after the pavement construction. This further supported the view that a conversion to gypsum occurred during construction, prior to stabilisation.

SUMMARY

The available information showed that:

1. The geological history and the mineralogy of the soil used for stabilisation had the potential to develop additional sulphate in the form of gypsum.

2. Within the earthworks excavation the construction conditions were suited to the oxidation of the sulphides present at an enhanced rate to form sulphates.

3. The addition of lime together with the soil mineralogy provided the environment for the formation of ettringite which was shown to be present.

4. Water beneath the structural pavement layers provided the water of hydration to the ettringite to initiate swelling. This was frequently evident where slack falls occurred in the sub-formation drainage or where changes in construction materials interrupted the direction of drainage.

5. Seasonal temperature changes in the field together with the soil mineralogy allowed the conversion to thaumasite, creating secondary gypsum providing a continuation of the process as long as the environment permitted.

The evidence which had been obtained on the Gaydon Contract indicated that the defective materials had suffered similar problems to those on the Banbury IV Contract.

OTHER CONSIDERATIONS

The considerable deterioration of the lime stabilised material into a very wet material with very low strengths required consideration to be given to the pavement design on the Banbury IV Contract. It was very fortunate that the design had incorporated a CRCR pavement construction and also a 600mm multi-element capping. In effect the construction still incorporated a 350mm rock capping on what had become a weak subgrade. This resulted in a non-standard Department of Transport pavement foundation. The loss of pavement life was determined and found to be sufficiently small to be offset by appropriate overlays at the first surfacing renewal, if the traffic growth predictions warranted this.

Levelling of the carriageways over the Banbury IV Contract continued at regular intervals, even after opening, until movements appeared to cease. This occurred in April 1992, 2 years after the first heave was discovered. Laboratory testing also indicated that it took about two years for the pH value of the stabilised material to reduce below 10.5, reducing the risk of further reaction. Figure 6 shows the decline in the rate of increase of heave over the two year period as an average of the values from the more significant areas.

IDENTIFICATION OF SWELL POTENTIAL

There are often savings to be made in construction costs by the use of lime stabilisation. On the other hand the rectification of distress caused by lime induced heave to a pavement can be extremely high, especially where reconstruction becomes necessary. With the awareness of the problem of lime induced heave the liability resting on the engineer has increased. Considerable care is therefore required in selecting a soil for lime stabilisation.

In the first stages of consideration a desk study establishing the geological history and mineralogy of the soil can identify a high risk soil. This can avoid the cost of extensive and expensive investigatory testing.

At the ground investigation stage sufficient testing to establish the total sulphur content, total sulphate content and the mineralogy of the soil is required together with the groundwater sulphate content. The inherent sulphate available to react with the lime and the potential for the production of additional sulphate by oxidation of sulphides can be assessed. If a laboratory testing programme to prove the latter is required a non-standard test such as that proposed by Hawkins and Pinches (1987) might be considered. Swell tests are unlikely to be of any value until all of the total sulphur in the soil has been converted to sulphate. Even at that stage it is difficult to reproduce the field conditions in the laboratory. Temperature changes together with water access need to be reproduced on the specimen. On the M40, swell tests on soaked specimens in CBR moulds proved to be unsatisfactory in reproducing either the degree of swell or the condition of the lime stabilised material in the field. The most significant test was carried out on a sample of field material mixed with lime, compacted into a cylindrical mould, and cured. The specimen was removed from the mould and immersed in water. Within minutes it began to disintegrate and within hours had collapsed completely.

If swell tests are required it is suggested that consideration be given to carrying out tests on cylindrical specimens which have been prepared to reproduce the stabilisation process. Curing at 20°C should allow the formation of ettringite if the environment permits. The specimens should then be placed upright in a water bath at 20°C. If it is felt necessary to provide confinement they could be supported by placing in a sand bed. Swell may be observed and at regular intervals the temperature should be cycled below 15°C and back again to observe the effects.

Full testing programmes to test sufficient material in order to obtain representative results can be expensive and time consuming. Therefore it may be better to assess the risk by desk study work or to minimise testing work to total sulphur and total sulphate tests and combining the results with a knowledge of the soil mineralogy.

The Highways Agency has recently produced a comprehensive advice note HA 74/95 covering the design and construction of lime stabilised capping for highway schemes which includes investigatory advice.

ACKNOWLEDGEMENTS

The author would like to thank Ove Arup and Partners and the Highways Agency for permission to produce this paper, and Dr D J Henkel for his significant contribution to the investigatory work.

REFERENCES

Caerns, P J and Noakes, R J (1988), "Lime Stabilised Capping at Saxmundham By-Pass (A12)", BACMI Symposium - Lime Stabilisation I C E.

Department of Transport (1976), "Specification for Road and Bridge Works", Fifth Edition, HMSO, London.

Department of Transport (1986), "Specification for Highway Works", Sixth Edition, HMSO, London.

Gallios, R W and Horton, A (1981), "Field Investigation of British Mesozoic and Tertiary Mudstones", Quarterly Journal of Engineering Geology, Vol 14, No 4.

Hawkins, A B and Pinches, G M (1987), "Cause and Significance of Heave at Llandaugh Hospital, Cardiff - A Case History of Ground Floor Heave Due to Gypsum Growth", Quarterly Journal of Engineering Geology, Vol 20.

Hawkins, A B and Pinches, G M (1987), "Sulphate Analysis on Black Mudstones", Geotechnique Vol 37, No 2.

Hunter, D (1988), "Lime-Induced Heave in Sulphate-Bearing Clay Soils", ASCE Journal of Geotechnical Engineering, Vol 114, No 2, p.150-167.

Mitchell, J K (1986), "Practical Problems from Surprising Soil Behaviour", 20th Karl Terzghi Lecture, ASCE Journal of Geotechnical Engineering, Vol 112, No 3.

Mitchell, J K and Dermatas, D (1990), "Clay Soil Heave Caused by Lime-Sulfate   Reactions", ASTM Symposium, San Francisco, Innovations and Uses for Lime.

Perrin, R M S (1971), "The Clay Mineralogy of British Sediments", Mineralogical Society (Clay Minerals Group), London.

Russell, D J and Parkes, A (1979), Quarterly Journal of Engineering Geology, Vol 12, p.107-116.

Shaw, H F (1981), "Mineralogy and Petrology of the Argillaceous Sedimentary Rocks of the UK", Quarterly Journal of Engineering Geology, Vol 14, No 4.

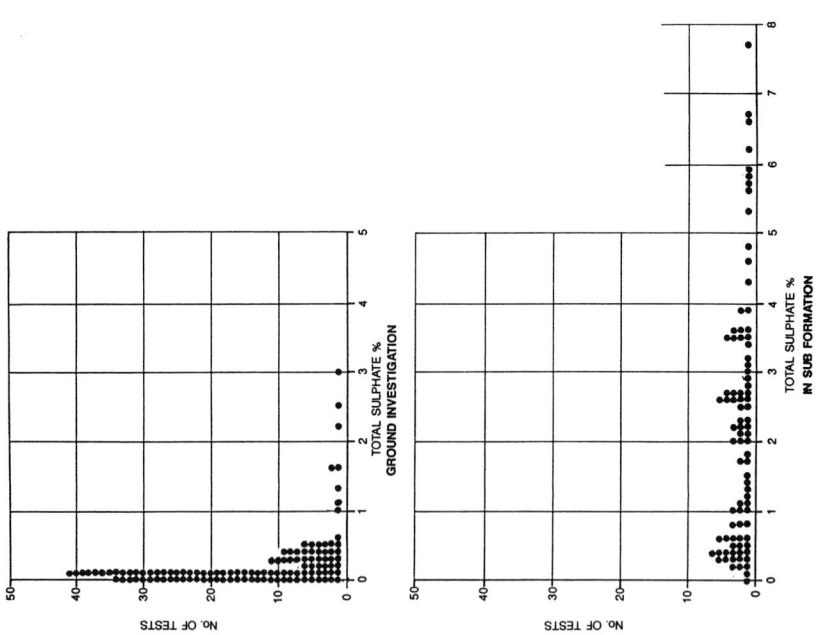

TOTAL SULPHATE % GROUND INVESTIGATION

TOTAL SULPHATE % IN SUB FORMATION

No. OF TESTS

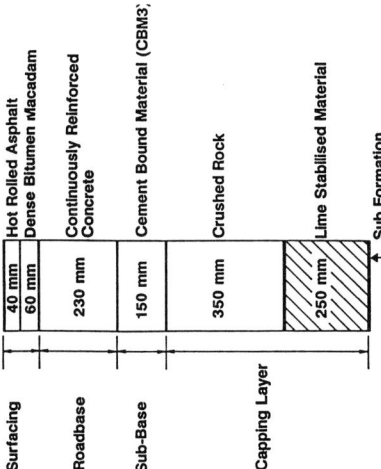

Surfacing
40 mm Hot Rolled Asphalt
60 mm Dense Bitumen Macadam

Roadbase
230 mm Continuously Reinforced Concrete

Sub-Base
150 mm Cement Bound Material (CBM3)

350 mm Crushed Rock

Capping Layer
250 mm Lime Stabilised Material

Sub Formation

Figure 1   Pavement and capping details Banbury contract

Figure 2   Total sulphate results Pre and post construction (Lower Lias)

154

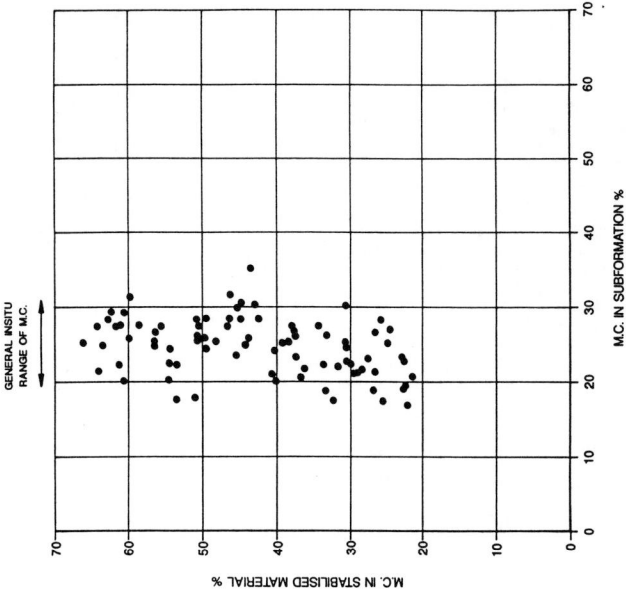

**Figure 4   Moisture content comparison
Stabilised and sub formation material**

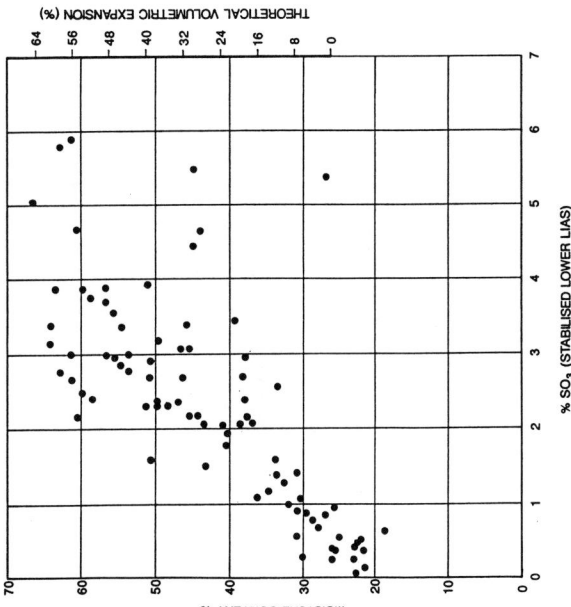

**Figure 3   Total sulphate v moisture content
Stabilised material**

155

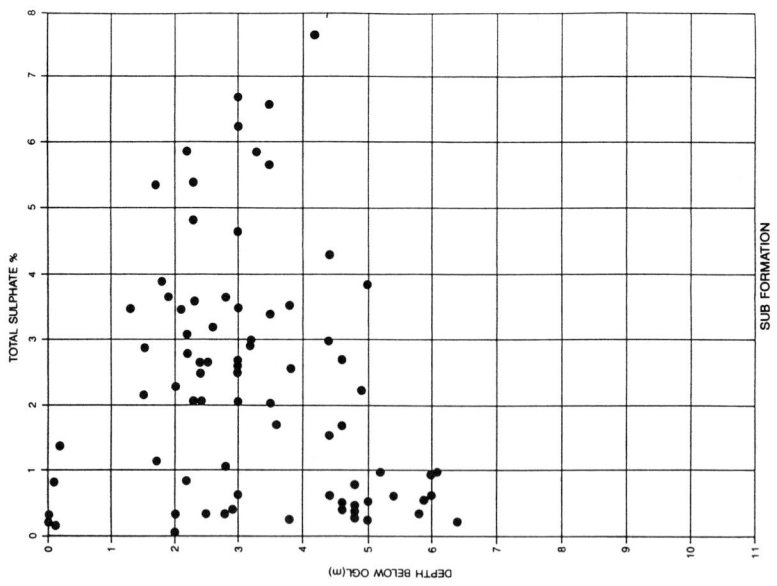

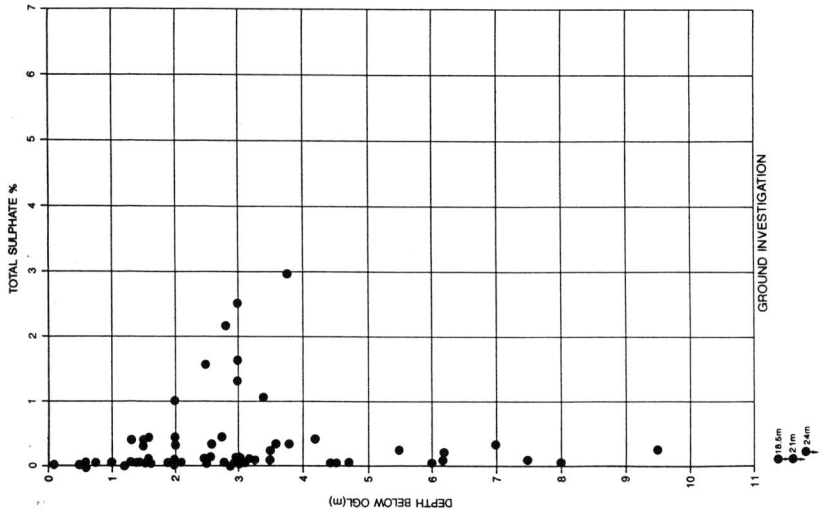

Figure 5   **Total sulphate v depth**
                      Pre and post construction

DEVELOPMENT OF HEAVE SINCE CONSTRUCTION

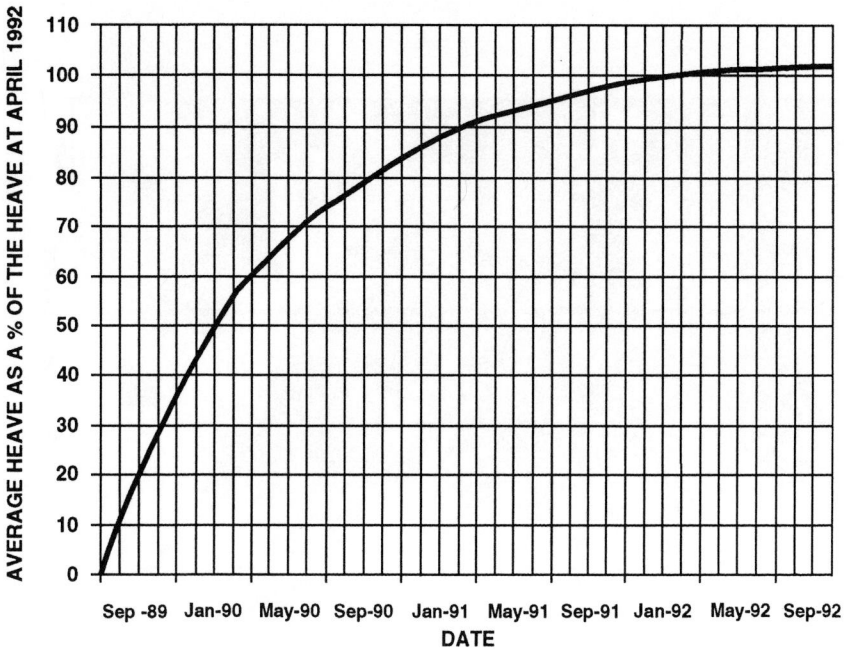

Figure 6   Results of level surveys
            Rate of heave

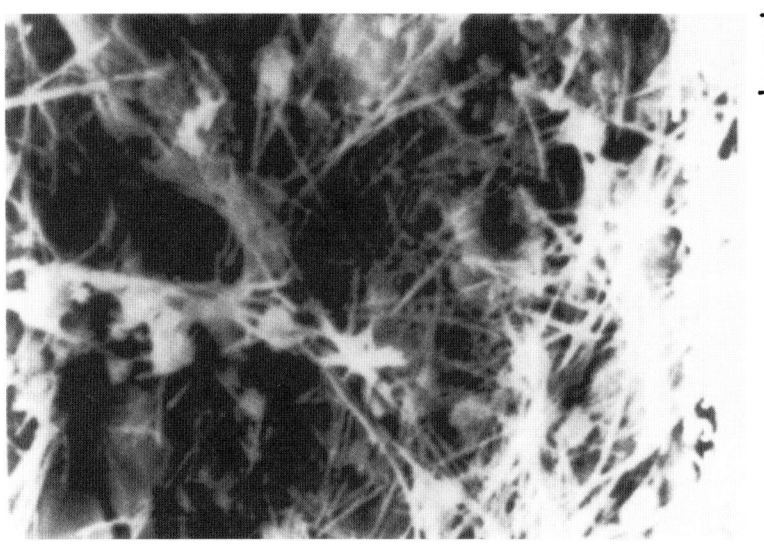

0.37% TOTAL SULPHATE   5.0% TOTAL SULPHATE

Plate 1   Scanning electron photomicrographs showing ettringite
           in stabilised soil specimens

158

Treatment of Silt using Lime and PFA to form Embankment Fill for the New A13

A. NETTLETON
Balfour Beatty Civil Engineering Limited

I. ROBERTSON
Hyder Consulting Limited

J. H. SMITH
Powerbetter Developments Limited

INTRODUCTION

The A13 (Thames Avenue to Wennington) is a new section of three lane highway which will by-pass Rainham and provide a replacement for the heavily congested existing A13 between the M25 and the Ford Motor Company plant at Dagenham. This contract is part of a larger scheme to provide a new or upgraded trunk road between the M25 and Canning town. The line of this contract runs across the Rainham marshes and through a series of silt lagoons (settling ponds), long used by the Port of Authority to dispose of silt dredged from the River Thames. The silt in the lagoons is up to 4.5m deep, with high moisture content, low shear strength and with some contamination. Consequently under the original scheme design it was required to be excavated and removed from site. The contractor for the works, Balfour Beatty, proposed an alternative which was to treat the silt so that it could be re-used.

The designer of the new highway on behalf of the Highways Agency, Hyder Consulting Limited, had to provide a scheme which would take into account the high level of settlement predicted by the ground survey. They approached this problem by using the following techniques;

1. Use of staged construction and settlement periods to allow the majority of settlement to take place prior to the installation of carriageway drainage and the road pavement.

2. Use of piles to support the high embankments adjacent to structures.

3. Formation of transition zones between the rigid piles and the remainder of the lower embankment which did not use any form of piled support.

One of the main elements of the design was the use of lightweight fill materials within the high embankments. Lightweight embankments would reduce the loading on the piles and provide a more economic solution. The lightweight fill adopted by the design was pulverised fuel ash (PFA), which has a bulk density between 1.4 and 1.6, whilst class 1 or 2 materials (granular or clays) have a bulk density in the region of 2 to 2.2. The use of PFA gives an overall weight reduction of approximately 25%.

DEVELOPMENT OF THE CONCEPT OF USING SILT AS A FILL MATERIAL

The idea of using the silt to fabricate a lightweight material came out of a necessity to solve a number of construction problems. The first problem was how to get rid of 300,000m^3 of a soft material with a moisture content in the region of 120-140%. The silt could not hold this amount of water, and great care would have to be taken to prevent any free water spilling out of the wagon, and onto the public highway. Removal by wagon was not permissible since the silt was contaminated with a number of chemicals found in the river Thames.

The second problem was the access into the site. The new A13 alignment runs over the main railway line linking Southend and London. This railway cuts the site in two and prevents access from the eastern end. The use of Bailey bridging techniques to span the railway were compromised by the restrictions on the adjacent land which formed part of a site of special scientific interest (SSSI). This left only one minor road as a permitted access onto the site. This road would therefore have to accommodate the removal of 800,000 m^3 of material and an import of 1,500,000 m^3 of material. This import requirement included 385,000 m^3 of lightweight fill for piled embankments. One way of reducing the affect of these wagon movements was to keep the silt on site and use it as lightweight fill material. If such an idea could be realised the wagon movements on and off site would be reduced by about 10% (40,000 No.).

As an idea, this had a number of shortcomings. The silt layer was very wet, approximately 4.5m deep, and sat on a layer of alluvial clay. The water level in the lagoons was 0.3m below the surface. Tracked plant working in the lagoons needed to sit on timber mats to spread the load and prevent sinking. However, when the need arose, a solution was found. In this instance, the idea came relatively early on in the contract, jointly developed by Balfour Beatty and Messrs. Powerbetter Limited.

The form taken by the early testing was to try mixing the silt (plus lime or ordinary Portland cement) with clay, PFA or special chemicals which were brought in from Holland, Switzerland and Japan. The composition of these chemicals was secret. This early testing indicated that a stable material could be obtained with both clay and PFA, plus mixtures of all three. The secret chemical was found to be workable, but a method of mixing the chemicals into the soft pliable silt could not be found.

The clay mixtures were ultimately dropped because of the high density of the mixture and the weaker physical properties of the blended materials. There was also the problem of finding a large source of quality clay which could be imported at a consistent rate. This left the PFA plus lime mixtures which provided the following advantages:

1. The resultant mixture had a density within the design parameters of the lightweight embankments and so could be substituted for the originally specified PFA.

2. The PFA was able to absorb a large quantity of water, and therefore use could be made of the water which came out as a by product of the silt. It should be remembered that water is required when placing and compacting PFA. This water prevents the PFA from blowing and also brings the water content of the material within the boundaries needed for acceptable compaction. The silt contained all the water needed, plus some excess.

3. The silt was lightly contaminated with chemicals picked up during its time in the river. The addition of the PFA effectively diluted the concentration of the chemicals, bringing the mixture within acceptable levels.

4. The PFA mixture produced a granular material with an internal angle of friction which complied with the design requirements for the lightweight fill.

5. The lime reacted pozzolanically with the PFA and silt, providing additional strength to the mixture, particularly in the longer term. It also reduced the high moisture content of the silt.

LABORATORY TESTING AND FIELD TRIALS

Before full scale field production could begin, it had to be shown to the client for the scheme, the Highways Agency, that use of the material would be technically acceptable. The treated material had to have acceptable strength and volumetric stability, and present no risk of pollution to the surrounding environment. A programme of laboratory and field trials was carried out to demonstrate that these criteria could be met, and to find the best method of treatment.

The potential problems posed by use of the treated silt were:

1. Need for high angle of shearing resistance for the material.

2. High moisture content, which had to be reduced from over 100% to around 40%.

3. High organic content, which could reduce effectiveness of stabilisation and degrade with time.

4. High sulphate content, with consequent risk of swelling.

5. Variability of silt, making it difficult to control moisture content for compaction.

6. Slight contamination, giving a potential for pollution of watercourses.

7. Environmental nuisance and health and safety hazard during treatment.

Further development and field trials

The early testing was carried out in the site laboratory and on small scale test sites adjacent to the silt lagoons. The aim of these tests was to provide enough information so that a submission could be given to the consultants and the Highways Agency. This submission would need a viable mix design which could be placed in large quantities and at high daily outputs. There were 300,000 m^3 of silt available for use. A slow process would not have been practical given the time available within the contract programme for the construction of embankments. Messrs Powerbetter were brought in at an early stage to give ideas on mix design. They had experience with lime stabilised materials and the techniques needed to place and compact the mixture. After several months of trials and testing the first large scale trial took place in September 1994.

This trial used both clay and PFA. Although the clay was not considered a long term option for this contract, the results could be useful should another contract require a material with different parameters. The plant and mixing techniques were provided by Messrs Powerbetter. This trial was not a complete success, although it did point out some of the shortcomings and give direction for workable solutions. The problems highlighted by this trial were as follows:

1. A PFA with a lower moisture content would have to be provided, or the moisture content and plasticity of the silt would have to be reduced prior to mixing with the PFA.

2. The thickness of the layers needed to be reduced. At this stage the mix consisted of silt:PFA at 1:1 by weight with approximately 3% lime. The layer depth used in the trial was 400mm. This proved to be too deep for the mixing plant to traverse with out heavy rutting, which it was desired to prevent.

3. The lime needed to be placed onto the silt before starting any mixing/rotovating of silt into the underlying PFA. The lime reacted with the water in the silt and also assisted in breaking up the silt into small lumps. Any mixing of silt and PFA prior to the addition of the lime was slow and problematical.

161

In spite of these problems, the trials showed that it was possible to produce a bulk fill material which had a granular consistency and complied with the two primary requirements of the lightweight fill, namely low bulk density and suitable internal angle of friction. From these aspects the trial was considered to be a success and a major step forward in the method of production.

After this trial, use of the treated silt was formally proposed to Hyder Consulting Limited and the Highways Agency. The Transport Research Laboratory (TRL) were also consulted. Further testing was required, which had to be independent and was therefore carried out at off site laboratories. This test regime is fully described below. The main thrust of the testing was to find out how the material would react in the long term and how it should be tested and monitored during production. Out of these tests came the detailed method statement and the specification for mixing and placing. This specification also included the types of test and the test frequency.

A second large scale site trial was carried out in March 1995, at which the Highways Agency were present. On this occasion Messrs Powerbetter had adapted the use of larger rotovation plant and the order of placing and mixing materials. This trial was a complete success, providing material even better than the laboratory trials had indicated. It was considered that the better quality mixing provided by the larger plant was providing the main increase in performance. Extensive compaction and laboratory tests were carried out on the treated silt from the second trial as described below.

The compaction method adopted in the second trial was to initially use a vibratory smooth wheeled roller. However only about 90% compaction could be achieved with this. A dead-weight sheeps pad roller was then used to achieve 95% compaction. If the sheeps pad roller was used in vibratory mode the material was loosened.

Laboratory testing

A range of laboratory tests was carried out on the treated silt, both on specimens prepared in the laboratory and on material taken from the field trials. The proportions of silt, PFA and lime were varied to find the range of mixes which could be successfully treated. The ratio of silt to PFA was varied between approximately 4:1 and 1:1 in the laboratory tests. Lime was added at amounts varying from 2.5% to 4% (expressed as a percentage of the wet weight of silt used in the mix). The consistency of the treated silt was that of a non-cohesive soil. The following types of test were carried out: classification; strength; compaction; moisture condition value; permeability; and swelling.

Properties of the untreated silt and PFA. Typical properties of the untreated silt are given in Table 1 below. Laboratory and field testing undertaken during previous ground investigation gave the following properties:

Table 1 Silt properties

Test	Unit	Range	Mean
Natural moisture content	%	54 to 159	118
Plastic Limit	%	40 to 80	56
Liquid Limit	%	83 to 159	129
Plasticity Index	%	32 to 92	67
Undrained shear strength (vane)	kN/m^2	5 to 12	-
Bulk density	Mg/m^3	1.23 to 1.55	1.42
Maximum Dry Density (2.5kg)	%	-	1.12
Optimum Moisture Content (2.5kg)	%	-	45
Total Sulphate as SO_4	%	0.54-0.73	0.60
Total Sulphur as S	%	0.06-1.20	0.82
Total Sulphur as SO_4	%	2.60-3.60	2.92
Organic Content (BS1377)	%	6.4-7.2	6.7

Particle size distribution analyses show the 'silt' deposit is actually a silty clay with a range of 35% to 70% clay size particles. The plasticity chart in Figure 1 shows the material is classified as an organic clay of high plasticity. The results of consolidated undrained triaxial tests on 38 mm diameter undisturbed samples are given in Table 2. The ø' of the silt is highly variable, reflecting the origin of the material and the variation in particle size.

Table 2 Effective stress shear strength - silt

Location	Depth	c' kN/m^2	ø' (°)
PS103	0.93m	0	44°
S4	3.57m	0	14°
S9	1.1m	5	22.5°

The organic content of the silt measured by BS 1377 was between 6.4% and 7.7%. Organic content testing by Toluene Extract gave a range between 0.3% to 1.15%, although some low boiling point organic material may have been lost. The proportion of Freon extractable organic matter was around 0.2%. The above indicates that most of the organic matter is not in the form of oils. A mass spectroscopy test confirmed that most of the organic matter is likely to be in the form of microscopic plant matter.

The PFA proposed for the silt treatment originated from stockpiles at Tilbury and Purfleet (formerly Thurrock). The properties of this PFA are given in Table 3. The PFA was imported to the site with a moisture content of around 25%.

Table 3 PFA properties

Test	PFA Source	
	Tilbury	Purfleet
Specific Gravity	2.25	-
Silt fraction %	50	-
Optimum moisture content (2.5 kg) %	23 to 37	30 to 36
Maximum dry density (2.5 kg) Mg/m^3	1.07 to 1.41	1.10 to 1.22
Effective angle of Friction, $\varnothing'$ ($^\circ$)	31 to 33	33
Cohesion c' kN/m^2	25 to 26.5	12
Total Sulphate as SO_4 %	0.62 to 1.36	1.00
Sulphur as S %	-	0.55
Sulphur as SO_4 %	-	1.70
Organic Matter (BS 1377) %	1.95 to 2.18	-

Compaction and moisture control of the treated silt. To ensure adequate strength and stability it was desired to compact the fill to greater than 95% of the maximum dry density, and to restrict the air voids to less than 10%. To achieve this the moisture content of the treated silt before compaction had to be controlled. This was particularly important due to the variable moisture content of the excavated silt. Compaction was investigated both in the laboratory, and on material compacted in the field. The aim was to establish the range of moisture contents and MCVs at which the above criteria were satisfied.

The conventional method of moisture control was first considered, whereby the allowable range of moisture content is related to the optimum moisture content (OMC). The dry density - moisture content relationship was investigated using the BS 1377 2.5kg compaction test, see Figure 2. The curves obtained were relatively flat, making it difficult in some cases to determine the OMC. The flat shape of the curve meant that 95% relative compaction could be achieved with the moisture content between OMC and OMC +/-7%. However it was desired to keep the moisture content greater than OMC, to keep the air voids low. Therefore limits of OMC to OMC+7% were considered suitable for the fill.

Compaction testing was carried out on laboratory mixed samples with mellowing periods of 2 and 24 hours. These showed that the OMC increased by 1-3% over this period, due to the action of the lime. For a mix of approximately 2 silt : 1 PFA at 24 hours, the OMC was in the range 46% to 53%, with a maximum dry density of 1.03 to 1.07 Mg/m^3. Compaction tests on samples of treated silt taken from the full scale field trials showed this material tended to have a lower optimum moisture content and higher dry density than the material prepared in the laboratory, see Figure 2. The OMC was in the range 30% to 40% and the maximum dry density was typically between 1.13 to 1.23 Mg/m^3. The reasons for this are not clear, but may be due to the difference in method and degree of mixing of the silt, PFA and lime between the laboratory and the field.

A series of compaction tests were carried out on samples from the field trial at different times to investigate the effect of mellowing. The results are given in Table 4. These show an increase in OMC between the test at 8 hours and at 2 days, but thereafter little change. In the full scale field production some of the material was pre-treated off-site, resulting in a mellowing period of several weeks. However there appeared to be no significant difference in OMC between this and the material with a mellowing period of 1-2 days. The maximum dry density of the material did not change significantly during the mellowing period in either the laboratory or field trials.

Table 4   Summary of laboratory testing on samples from the field trials

Time elapsed since adding lime	Bulk Density (Mg/m^3)	Optimum Moisture Content (%)	Dry Density (Mg/m^3)	MCV at Optimum
8 hours	1.57	30	1.20	19
2 days	1.64	40	1.17	13
4 days	1.61	36	1.19	14
8 days	1.63	37	1.19	17

The change in OMC with time and method of mixing made it difficult to control moisture content for compaction by relating moisture content to OMC. Instead it was decided to try and use the Moisture Condition Value (MCV) test to control moisture content. This test has the advantage of self- compensating for the changes with time and method of mixing. It would also give a rapid indication of whether the treated silt was ready for compaction, or whether further treatment was required.

In order to use the MCV test, a correlation of MCV with moisture content had to be established. This was again done by laboratory tests on material mixed both in the laboratory and in the field trials. The laboratory mixed material indicated that an MCV of between 8 and 13 would be suitable after a 24 hour mellowing period, corresponding to OMC to OMC +7%. However as for the optimum moisture content, the MCV correlation for the material mixed in the field was different to the material mixed in the laboratory. The tests on the material mixed in the field indicated a suitable MCV range of 10 to 15. This MCV range was confirmed by correlating directly the MCV of material before compaction with the dry density achieved *in situ* in the early stages of field production.

It had to be ensured that the degree of mixing achieved by the treatment was acceptable, and that there would be no pockets of material with a moisture content outside the acceptable range for compaction. Visual assessment of the material in the field trials indicated that a high degree of mixing was achieved. Moisture content tests taken at random from the field trial confirmed that it was acceptable.

In summary, a method had to be found which would enable the moisture content of the treated silt to be controlled, to ensure that it could be compacted to 95% of the maximum dry density. The optimum moisture content of the material was difficult to determine and tended to change with time and method of mixing. The MCV test offered a quick and reliable alternative method of control. The range of MCV was determined from testing material produced and compacted in the field trials.

Strength. The effective stress shear strength parameters of the treated silt were determined by both drained shear box (60 mm) and consolidated undrained triaxial tests. The triaxial tests gave an average ϕ' of 29°, and a minimum value of 28°. The triaxial test is considered to give a lower bound strength of the material, as it is fully saturated. The effective cohesion c' obtained from the triaxial tests was in the range of 11 to 34 kN/m^2, with a mean of 20 kN/m^2. Although the ϕ' of the material was slightly less than the required value of 30°, it was considered acceptable due to the high cohesion. The shear box tests gave an average peak ϕ' of 33°, with a minimum of 29.5°. The residual strength of the material was between 1° and 2° lower than the peak. The strength of the treated silt did not vary significantly within the range of mixes tested. It was concluded from the tests that the strength of the material was acceptable, provided 95% relative compaction was achieved. Shear box tests were specified to be carried out on the treated silt during full scale field production to monitor the strength achieved.

Swelling, sulphates and organic content. There was concern that the treated silt should have acceptable volumetric stability, particularly as the silt had a high sulphate content. As discussed above, it was considered that compaction to 95% relative density and restricting air voids to less than 10% would limit this risk. A series of eight 28 day swelling tests were carried out in a CBR mould on material prepared in the laboratory and taken from the field trials, compacted to 95%. In all cases the recorded swell was <1 mm.

The results of sulphate and organic tests on a mix of approximately 2 silt : 1 PFA with 5% lime are given in Table 5 below. The acceptable upper limits of sulphate content for the treated silt were determined from the sulphate content of samples which had acceptable swelling properties.

Table 5   Sulphate and organic content of treated silt

Test	Standard	Range	Mean
Total Sulphate as SO$_4$ %	BS 1377	0.7-1.0	0.9
Total Sulphur as S %	BS1047	0.5-0.9	0.7
Total Sulphur as SO$_4$ %	BS 1047	1.6-2.6	2.2
Organic Matter %	BS 1377	4.7-9.6	6.0

The organic content of the treated silt remained high with a mean value of 6%. This is in excess of the upper limit of 2% recommended for stabilisation of cohesive material to form a capping in the DoT Specification for Highway Works. However as the swelling properties were within limits, the organic matter did not interfere with the lime treatment, and the required degree of compaction could be achieved, the organic content was judged to be acceptable. It is considered that the breakdown of organic material with time due to anaerobic biological degradation will not be a significant problem. This is because the treated silt is effectively sealed in the core of the embankment, and little water will be available. Also the treated silt is likely to be highly alkaline, which would inhibit breakdown. Monitoring of the internal settlement of the embankment fill is being carried out to check if there is any degree of degradation.

Permeability, chemical and leachate tests. The main environmental concern of using the material, apart from dust during construction, was the potential for production of leachate which could pollute surrounding watercourses and groundwater. This problem was investigated in two ways:

chemical testing of the treated silt and leachate produced from it;

consideration of likely infiltration and run-off using measured permeability.

Chemical testing on the treated silt for a wide range of determinants showed it to fall into the class of 'slightly contaminated' as defined by HSE Guidance Note HS (G) 66, due to the levels of mercury, cadmium, chromium and zinc. The levels were below the ICRCL trigger values. Leachate tests on the treated silt (approx. 2 silt: 1 PFA with 3-5% lime) showed that the leachate met the Water Quality Standards for Irrigation. The lime provided a considerable amount of cementing due to pozzolanic reaction between the lime and PFA, and increased pH. This greatly assisted in the low permeability and the reduced solubility of metals.

Constant head permeability tests were carried out on samples of the treated silt, including 4 on material prepared in the laboratory and 1 on material taken from the field trial. The ratio of silt : PFA was varied from approximately 4:1 to 1:1 in the laboratory prepared samples, see Table 6.

Table 6 Permeability test results

Mix ratio (silt: PFA) approx.	% lime	Permeability m/s$*10^{-8}$
4:1	3.0	0.8
2:1	2.5	0.8
2:1*	4.0	6.3
1:1	3.9	1.1
1:1	3.9	2.2
Neat PFA	-	2.3
*sample taken from field trials		

The permeability values of the treated silt were all low and were comparable with neat PFA. There was a slight tendency for permeability to increase with reducing proportion of silt. The maximum value of $6*10^{-8}$ m/s was obtained on the sample from the field trial. The treated silt is virtually encapsulated beneath the road pavement, and the landscaping fill on the outer shoulders of the embankment. This coupled with the low permeability of the material, means that the rate of seepage through the fill will be very low. Consideration of the chemical and permeability data indicated that there was no potential for pollution as a result of using the treated silt.

In summary, the field trials and laboratory testing showed that the treated silt could be produced under site conditions and had acceptable engineering properties. A method statement and specification for the work were drawn up, and a departure from the DOT Specification for

Highway Works was sought. The sequence of treatment and compaction was further refined during the early stages of the full scale production work.

PRODUCTION AT FULL SCALE

Production at full scale took place between May and September 1995. A total of approximately 100,000m^3 of the silt was treated to produce around 150,000 m^3 of light weight fill. In order to maximise production and to minimise the environmental nuisance, much of the material was treated in an area adjacent to the site, which was leased for this purpose. After treatment the material was excavated, taken to its destination in the works and then rotovated and compacted.

Method

Most of the silt was treated in a ratio of 2:1 (silt : PFA), with the addition of 3% lime. The method finally adopted for the treatment of the silt is summarised below:

1. A 100mm layer of conditioned PFA was spread by dozer

2. A 200mm layer of silt was then spread on the PFA using a backactor

3. Lime was spread on to the silt and then mixed in using plough and harrow

4. The mix was then rotovated with a minimum of two passes

5. MCV was tested and if acceptable the treated silt was then compacted with a few passes of a smooth wheeled vibratory roller to seal the layer

6. If MCV was too low, more PFA was added followed by further rotovation

7. Final compaction was carried out with 8 passes of a dead-weight sheeps pad roller

8. In-situ dry density was checked by core-cutter method

Specification

A specification for the material had to be prepared, as it was a departure from standard and was not covered by the DOT Specification for Highway Works. The specification was end-product, requiring the dry density to be 95% of the maximum dry density as determined by BS1377: Part 4 (2.5kg rammer). In addition, limits on the following properties were specified, based on the results of the trials, and the requirements for the design of the permanent works:

  mix proportions (lime, silt, PFA)
  MCV before compaction
  bulk density
  effective shear strength (peak and residual)
  sulphate
  organic content
  swell at 28 days

Testing was specified during construction to control the above properties, which was carried out jointly by the Contractor's and the Engineers' site laboratories.

Compaction

The specified 95% compaction was achieved. The in-situ dry density is plotted against moisture content in Figure 3. The plot shows that the majority of the fill has air voids within the range of 5% to 10%. The maximum dry density achieved was around 1.18 Mg/m^3. A range of laboratory compaction curves on the field mixed material is shown for comparison. These confirm that most of the material was compacted wet of optimum. Figure 4 shows a histogram of the excess moisture content at the time of compaction in relation to the optimum moisture content (mc-omc). This indicates that 95% compaction was achieved with moisture content up to 15% in excess of the optimum, indicating a very flat compaction curve. This is in contrast to the trial laboratory results which indicated an upper limit of OMC+7%. The difference in the behaviour of the field and laboratory materials, and the difficulty of using OMC as a control is highlighted by these results.

MCV is plotted against relative compaction in Figure 5. The plot includes the areas which did not achieve 95% compaction first time and had to be re-worked and re-compacted. The plot shows that the MCV is particularly useful for predicting the upper limit of moisture content which can be compacted to 95%. The majority of areas which were compacted with an MCV of less than 10 did not achieve 95%. In these areas further PFA was added to the treated silt to increase the moisture content before re-compaction. Where the MCV was in the range of 10-15 prior to compaction, the majority (75%) of in-situ densities were 95% or more. In the areas which did not achieve 95%, further compaction was carried out, until it was achieved.

Where the MCV was in excess of 15, satisfactory compaction was achieved in the majority of cases. The concern at high MCV was that the material would have significant air voids if compacted too dry, giving a potential for swelling or collapse settlement. However the moisture content of the material with MCV >15 was still wet of optimum in most cases. MCV is plotted against air voids in Figure 6. This showed that an air voids content of greater than 10% only occurred where MCV was in excess of 12. However even where MCV was greater than 12 the majority of air voids were below 10%. This indicates that if a very tight control on air voids had been required, then an MCV range before compaction of 10 to 12 would be suitable. However a reasonable level of control is achieved with an MCV range of 10 to 17.

The relationship of MCV with moisture content from the field production is shown in Figure 7. There is a high degree of scatter, although the trend can be seen of increasing moisture content with decreasing MCV. The regression line through the points is shown, together with a regression line from the laboratory MCV calibrations. The difference in behaviour between the field mixed and laboratory mixed material is again clearly shown.

Engineering properties

The engineering properties of the treated silt as determined by the testing during the field production are shown in Table 7. The properties were all within the specified limits.

Table 7 Engineering properties of treated silt

Property	Units	Mean Value	Range
bulk density	Mg/m^3	1.55	1.41-1.64
maximum dry density (2.5kg)	Mg/m^3	1.14	1.03-1.25
optimum moisture content	%	32	24-39
effective cohesion c'(peak)	kN/m^2	24	13-38
effective angle of internal friction ø' (peak)	(°)	34	30-38
effective cohesion c'(residual)	kN/m^2	25	16-32
effective angle of internal friction ø' (residual)	(°)	31	30-32
swelling at 28 days	mm	1.4	<1-3.9
specific gravity	-	2.3	2.19-2.36
total sulphur	%	0.41	0.28-0.55
total sulphate	%	0.74	0.32-1.10
organic matter	%	4.6	1.9-6.8

Monitoring of internal settlement of the fill

It was considered prudent to monitor the internal self-settlement of the treated silt fill in the embankments. For this purpose magnet extensometers were installed in the embankments in some areas, with settlement plates within the fill material. Monitoring to date has shown up to 1% internal compression of the fill as the embankment was raised. Post construction internal settlement of the embankments has been around 0.2% to date. Monitoring is being continued, but the results to date are considered to be acceptable and confirm that significant degradation of organic matter is not occurring.

CONCLUSIONS

1. The material formed a stable base for running plant on.

2. There were significant differences in behaviour between the material mixed by hand in the laboratory and that produced by machine in the field. This highlights the need for a full scale field trial as part of the development of the treatment process.

3. The Moisture Condition Value (MCV) test was useful in controlling moisture content of the treated material prior to compaction.

4. The material appears to be stable with time. Some of the material was dug-up and re-compacted up to 1 year after treatment with no significant change in properties.

5. Use of the treated silt offered considerable environmental benefits, by reducing import of material to site and export of contaminated silt to landfill.

6. The treatment of the silt with lime and PFA to form embankment fill for the A13 has been a success.

ACKNOWLEDGEMENTS

The assistance of the Highways Agency and the Transport Research Laboratory during the development of the proposal to treat the silt are gratefully acknowledged.

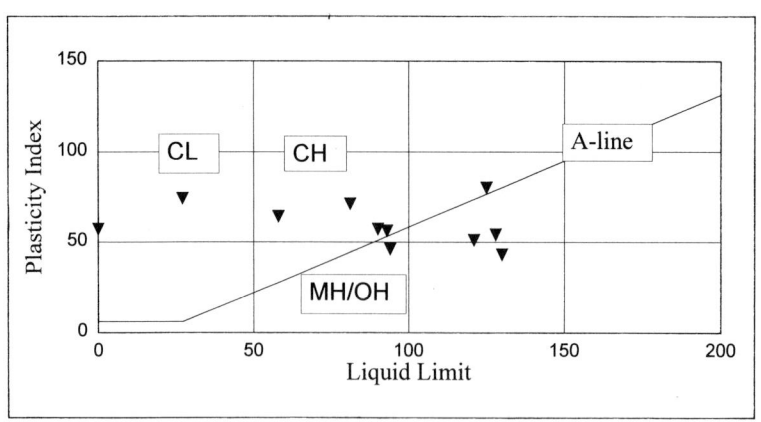

Figure 1 Plasticity chart for treated silt

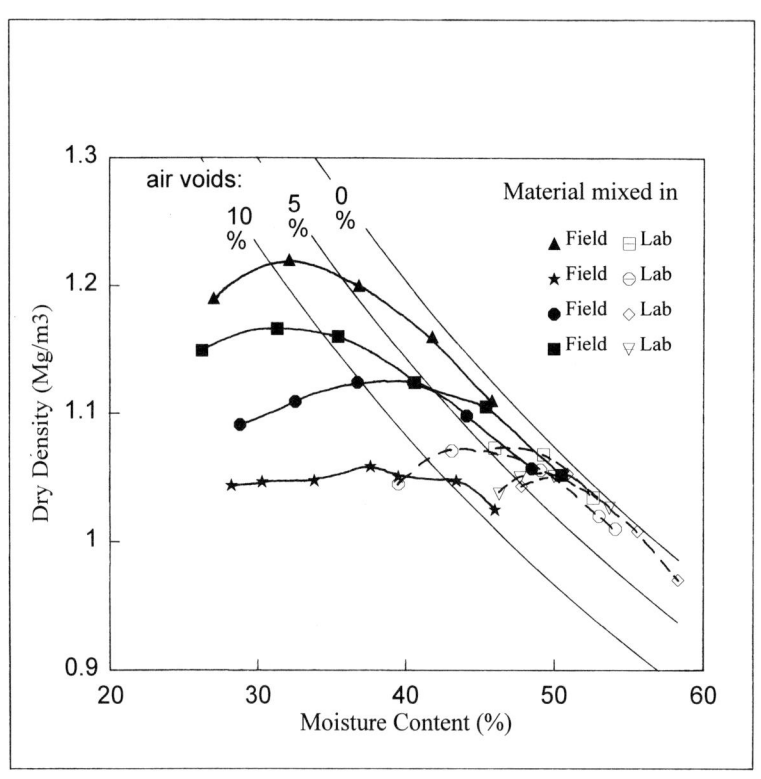

Figure 2   Treated Silt - Comparison of compaction curves from field and laboratory mixed material

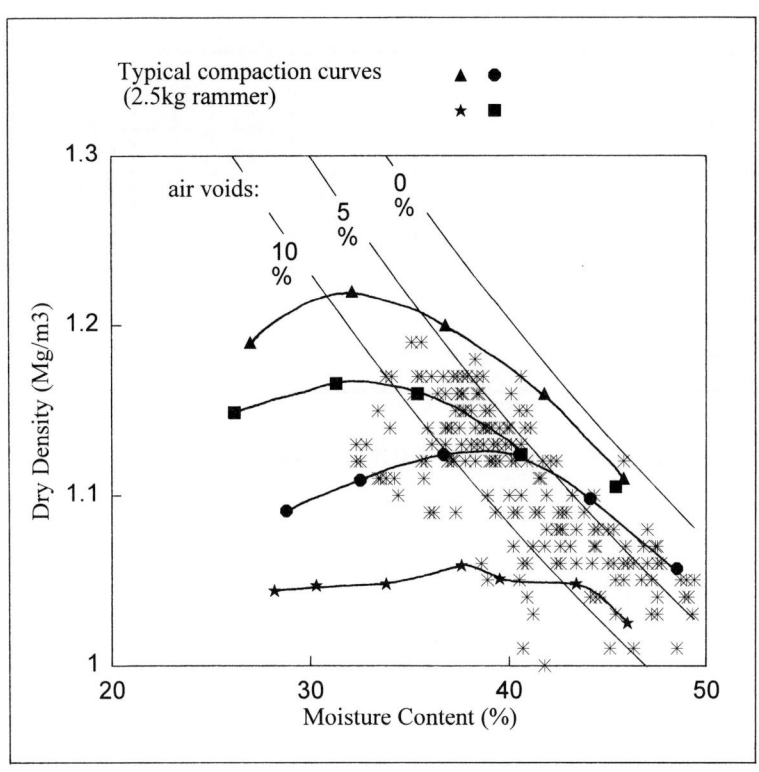

Figure 3   Treated Silt - In-situ dry density v moisture content

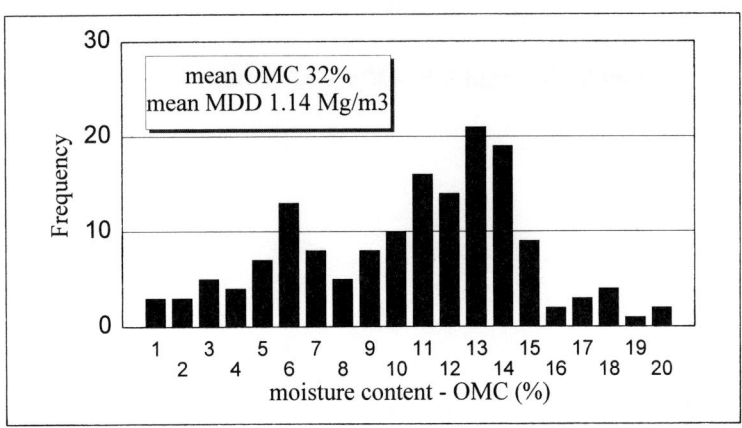

Figure 4   Treated Silt - Moisture content in excess of optimum at compaction

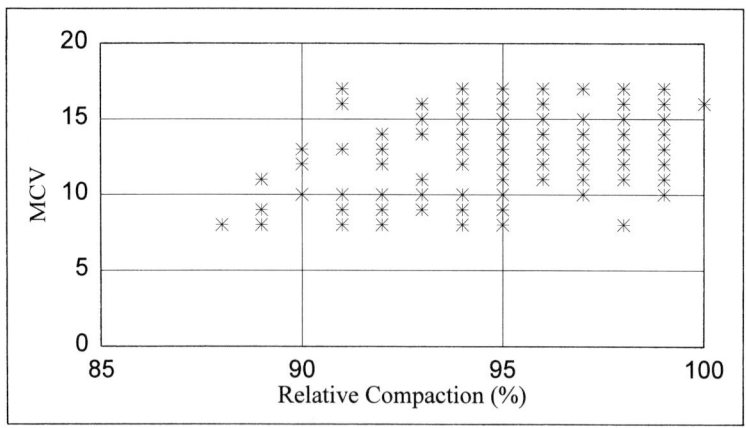

Figure 5   Treated Silt - MCV v Relative Compaction (%)

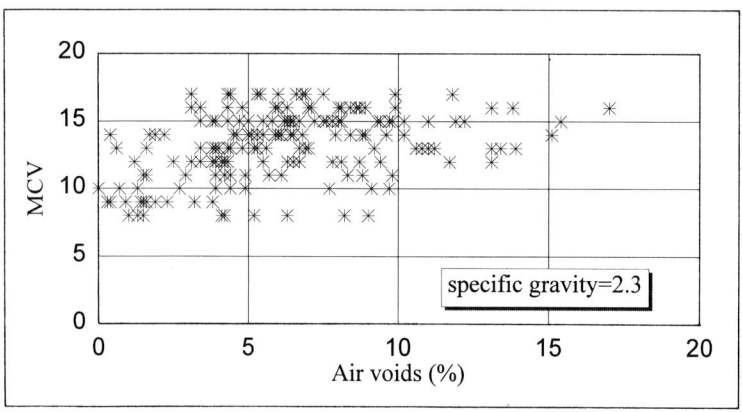

specific gravity=2.3

Figure 6   Treated Silt - MCV v Air Voids (%)

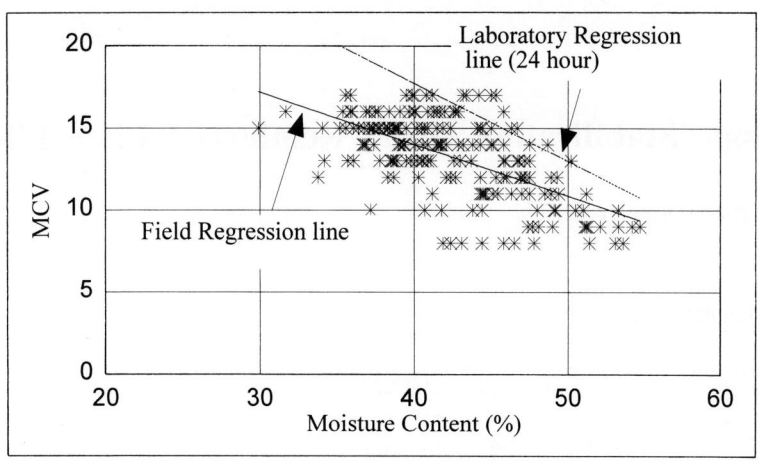

Figure 7   Treated Silt (field mixed material) - MCV v Moisture Content (%)

Slope Stabilisation Using Reinforced Lime Piles

L.THREADGOLD
Geotechnics Limited

INTRODUCTION

Most commonly, slope failure is recognised by the formation of a scarp face at the back of a slip, the accumulation of soil or rock at its toe and distortion or cracking of the ground surface between. Discrete failure surfaces or zones are formed between the sliding mass and the more stable ground beyond. For some embankments, however, the problem can be more subtle and more severe since movements of slopes with Factors of Safety against slippage which are low, but above unity, can render rail tracks or road surfaces unserviceable.

The solutions to both problems, used either individually or in combination, include:

1. Structural support provided by a retaining structure or anchorage.

2. Adjustment of slope geometry by reducing the slope angle or introducing a berm to counter the imbalance of forces which led to instability.

3. Increasing resistance to slippage often addressed by the introduction of drainage and/or higher shear strength soils, across the shear surface or zone.

Slopes adjacent to highways or railways frequently pose access difficulties and place constraints on the remedial works options which are available. Trees, shrubs and vegetation contribute to stability and hence there is significant benefit in a solution which allows them to be preserved, minimises disturbance, selectively treats critical zones, increases the strength of the soils themselves and introduces resistance across the shear zone. Lime piles have been used overseas for this purpose but their use in this country is only now beginning to be recognised. This paper describes their function and a development in the UK which arises from observations of their behaviour.

CONCEPT

The use of lime essentially addresses the slope stability problem by increasing soil strength. In concept it does this in several ways:

1. The introduction of quicklime creates a demand for the water which is present in the clays and hence:

   a) Reduces pore water pressure by applying suction which increases effective normal stress and hence shear strength in the shear zone.

   b) Consolidates the shear zone.

   c) Increases undrained shear strength as a result of a reduction in water content.

2. Introduces a material which gains in strength as it absorbs water.

Lime stabilisation. Thomas Telford, London, 1996

3. Modifies the clays around the pile and increases their shear strength.

In order to enable practical implementation of this technique it is clearly necessary to quantify the benefits which arise from the various phenomena so that a design strategy can be formulated, the intensity, depth and location of treatment can be evaluated and the costs determined. This is discussed in more detail by Glendinning and Rogers (1996) in Chapter 3.

APPLICATION

In forming a new embankment is it likely that the criteria for acceptability of soils for this purpose should ensure that the required profiles would have an adequate Factor of Safety against shear failure. Where steeper slopes are required, soil reinforcing techniques can be used, unsuitable soil could be rendered suitable, by mixing it with lime for example, or granular layers can be introduced to provide both drainage and enhanced shear strength in cohesive soils. Hence, the use of lime piles or columns is unlikely to be appropriate to the formation of new embankments.

In new cuttings, lime piles could be considered as one of several options to create a slope steeper than would otherwise be appropriate for such materials. On a construction site, however, the problems of access are normally not severe and the use of the technique will depend upon its economy relative to a wide range of alternatives.

For existing slopes where instability is occurring, however, the use of lime piles can have significant benefits over the other options. Delayed failures are typically associated with slopes in clay since their stability tends to be greatest immediately after formation and reduces with time as shear strength equilibrates to ambient stress conditions and/or pore water pressures. Lime provides particular benefits in such soils through modification and the suctions which its demand for water creates, particularly seeking the higher water content soils often associated with the shear plane or zone. For many slopes which are moving, it may be necessary to increase the Factor of Safety by only a small amount to arrest movement. Hence large increases in Factors of Safety may not be required.

MECHANISMS

Suctions

The levels of suction which can be achieved by placement of lime piles in clays has been researched by Rogers and Glendinning (1993) and pore suctions of up to 20 kPa have been measured. In effective stress terms this can be perceived as enhancing the cohesion intercept (c') by an amount equivalent to the product of pore suction (U_S) and the tangent of the angle of internal shearing resistance (tan ϕ'). This effect is particularly significant at shallow depths where strength is dominated by the c' value. At depth the proportional increase is less but the effect nevertheless can be significant. The lime acts as a "pump" to pull out water from the clays in a manner which neither simple pumping nor gravity drainage could achieve. Its effects can be very rapid and arrest movement in an unstable slope since it particularly targets the wetter zones or shear planes associated with instability. Such pore water pressure reductions have the beneficial effects of surcharge, by consolidating softened zones, without the de-stabilising down-slope components of conventional surcharging. Such consolidation is likely to lead to enhanced shear strength both in the short term and long term.

Shear strength of a lime pile

There is concern that suctions or pore water pressure reduction may not continue in the long term unless measures to exclude water are also taken. It should be noted, however, that as the tendency to create the demand for water reduces, so the lime gains in shear strength. Work in

Japan with lime (Kitsugi and Azakami, 1982) has shown that shear strengths of the order of 200 kPa are achieved by the lime as it becomes hydrated *in situ* . Work by Rogers and Glendinning (1993), has measured strengths of lime produced in the UK of up to double this value so that 200 kPa would seem to be a reasonably conservative value to take for design purposes at this stage. Again the effect is to increase the shear strength across the shear plane.

Shear strength of clay

Work overseas has claimed major improvements in the clays adjacent to the lime piles as a result of physicochemical reaction. Work in the UK (Rogers and Glendinning, 1993) indicates that such benefits only extend for some 30mm into the clays, however. Hence, this effect may not be as great as perceived, although a 30mm increase in the radius of a small diameter pile can be significant. Secondary effects resulting from clay desiccation and the creation of radial "fissures" filled with lime may result in the influence being greater than 30mm in zones near to such fissures. The consequence of this chemical reaction is to create higher strength at a given water content and effective stress condition. This effect, together with the lime pile increase in strength, is long term and not as potentially transient as pore suction.

The effect of the lime piles will of course depend upon their spacing, diameter and length. In the short term it will influence suctions and in the long term the gain in shear strength of the lime-clay system.

DESIGN

A substantial number of analyses of slope stability have been conducted by Lucas (1996) to evaluate the depth to which lime pile treatment is most effective and the zone in the slope in which it would create the greatest benefit. For the purpose of this study, Bishop's method of analysis for circular potential failure surfaces was used. This was implemented on computer using software (BISHOP) developed by Bromhead (1985). These analyses showed that the effect of treatment is most significant at shallow depth (Figure 1) but since most failures of slopes are less than 2m deep (Perry, 1989) the treatment is appropriate for a high proportion of slips in the UK. Deeper treatment can be effective in raising Factors of Safety for deeper potential slips to above the target figure, even though the proportional increase is small. For example a Factor of Safety for a shallow slip may be raised by 30% from 1.0 to 1.3 whilst a deeper seated potential failure surface having a Factor of Safety of say 1.25 requires only a 4% increase to raise its Factor of Safety to the same value.

In relation to the location of treatment, it would appear that in most instances in a landslip, movement tends to be initiated in the lower third to half of slope width. Treatment in this zone and other parts of the slope were modelled in the analyses but no significant trends became apparent. This is likely to be due to the modelled benefits not being largely stress dependent. Consideration of the mechanisms of failure, however, suggests that the treatment should be downslope of the neutral point of the potential failure surface. This would tend to create a buttressing action to the remainder of the slope and be in the zone of highest local shear stress where movements are typically initiated. Piles can most readily be put in vertically but they can also be installed at an angle to the vertical in order to target particular zones from accessible locations on the slope.

DEVELOPMENTS

It is known that lime tends to expand on hydration and in an unconfined state, expansions of 100% have been recorded. In the field, evidence of this has been seen in some of the piles in which lime exuded out of the top of the hole. As confining pressures increase so the tendency to expand decreases although simple consolidation tests in an oedometer have shown that to prevent expansion, very high loads have to be applied (Greenwood - private communication).

178

This observation led to the development of a patented design shown in Figure 2 in which a central anchor bar with fixed end plates is installed in association with the lime. Hence, any tendency of the lime to expand axially is restricted with the result that the lime is forced into contact with the adjacent clay. This enhances the bond between the lime and the clay, puts the central rod into tension and hence works together with the lime to enhance shear strength. It also provides an anchor with tensile capacity to further increase the Factor of Safety for the slip and to render the technique suitable for use at higher levels in a slip as well as in the lower parts. The combined clay-lime-steel system is designed to create a much strengthened and stiffened zone (Figure 3).

EXAMPLES OF UK USAGE

Three major commercial contracts which make use of these techniques have been undertaken by Cementation Piling and Foundations Limited to treat embankments. These embankments were formed in the 1920's and 1930's to carry London Transport trains on embankments in the London suburbs. Little attention was paid to the compaction of the London Clay and other clays from cuttings and tunnels at that time, leaving a legacy of embankments which have subsequently given rise to slippage and on-going deformation. This creates major maintenance problems for the tracks and causes disruptions and speed limitations for the trains. The problems are not confined to the clays in these situations, since the ash which was used to infill areas of subsidence in the past has also moved. The ash has been stabilised using mix-in-place grout-ash "logs", to form a type of in-situ crib walling, or a piled and anchored or propped wall. Berms have also been used to enhance stability at some locations. The clay slopes or shoulders have been treated with reinforced lime piles. On the first site (Figure 4), lime piles were installed in the lower part of the slope to depths of up to 7m to increase the Factor of Safety for shallow slips and to constrain potential failure modes to depths below which the target Factors of Safety were exceeded. At some locations horizontal reinforced lime "piles" or nails were taken through the embankment to treat the clays which were close to track level. They not only provided stability by virtue of their anchoring effect but also stiffened clays beneath the ash.

At two further sites (Figure 5), the treatment using lime has been applied to the embankment shoulders in clay in order to stiffen them and hence reduce movement. Here they form part of the overall slope stability enhancement scheme, the combinations of which depend on the ground conditions at each location.

CONCLUSIONS

As work has progressed, techniques for installation of lime piles, in combination with reinforcement, have evolved into an efficient process. The economies which result from such treatment can be substantial as the work in the London Underground embankments has shown (New Civil Engineer, 1995; Ground Engineering, 1996). They provide a useful part of the armoury available to the engineer who wishes to address the problem of slope instability and can be used without obviating other solutions. They are particularly useful in the treatment of existing failing slopes in clay which are becoming ever more evident on the cut slopes in motorways and, less obviously, on embankments where the consequences of failure can be even more severe.

Bearing in mind the relative novelty of the technique, monitoring of its behaviour and effects is critical for a better understanding and more efficient application of this method of slope stabilisation. Research and development of this technique is continuing.

REFERENCES

Bromhead, E N (1985), "Manual for Bishop's Method of Slope Stability Computer Program".

Glendinning, S and Rogers, C D F (1996), "Deep Sabilisation using Lime", Lime Stabilisation, Edited by C D F Rogers, S Glendinning and N. Dixon, Thomas Telford, London.

Ground Engineering (1996), June, p.26-27.

Kitsugi, K and Azakami, R H (1982), "Lime-Column Techniques for the Improvement of Clay Ground", Symposium on Recent Developments in Ground Improvement Techniques, Bangkok, p.105-115.

Lucas L, (1996), M.Phil Thesis, in preparation.

New Civil Engineer (1995), 22nd June 1995, p.18-19.

Perry, J (1989), "A Survey of Slope Condition on Motorway Slopes in England and Wales", Research Report RR199, Transport Research Laboratory, Crowthorne, Berks.

Rogers, C D F and Glendinning, S (1993), "Stabilisation of Embankment Clay Fills using Lime Piles", Proceedings of the International Conference on Engineered Fills, Newcastle upon Tyne, September, Thomas Telford, p.226-238.

FOS IMPROVEMENT SLOPE 1:2
AGAINST SLIP MAX DEPTH

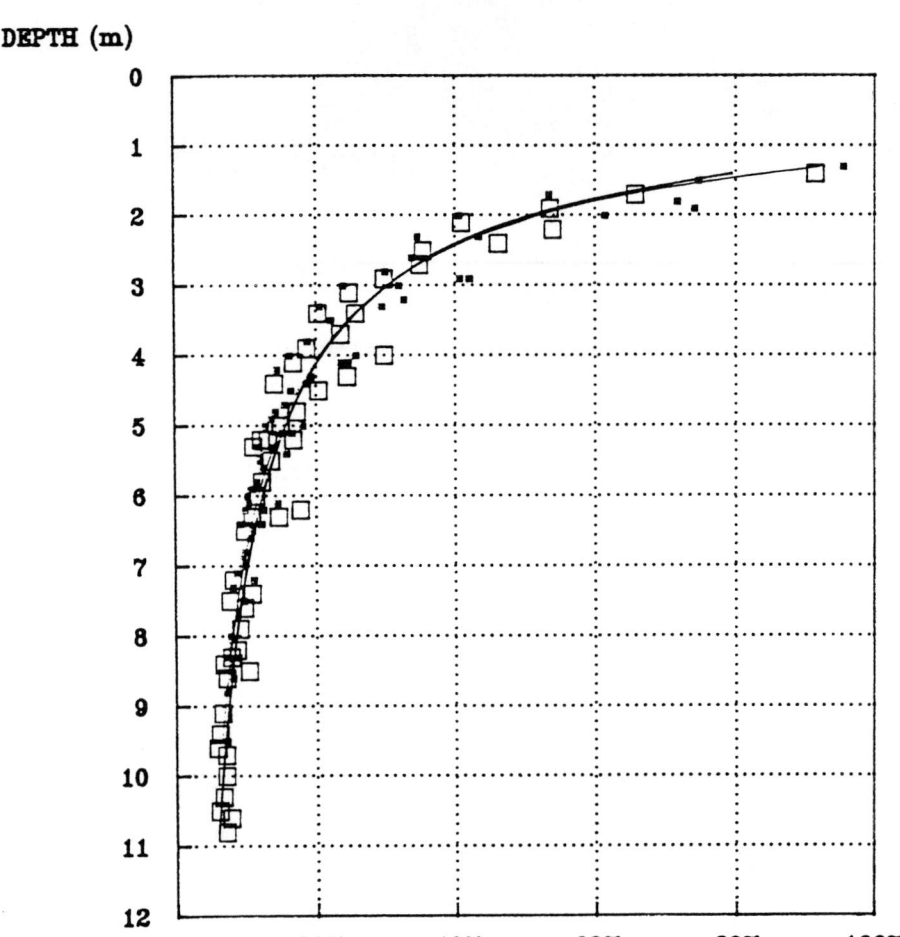

Figure 1   Degree of improvement in Factor of Safety with depth of
treatment.

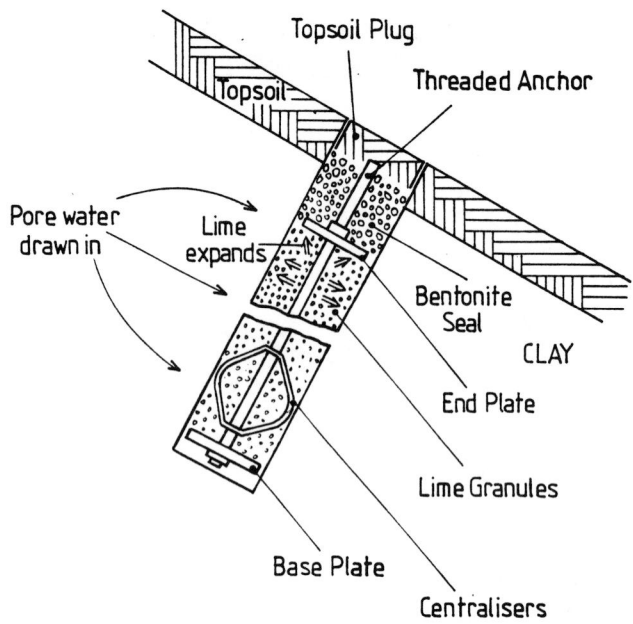

Figure 2   Detail of reinforced lime piles.

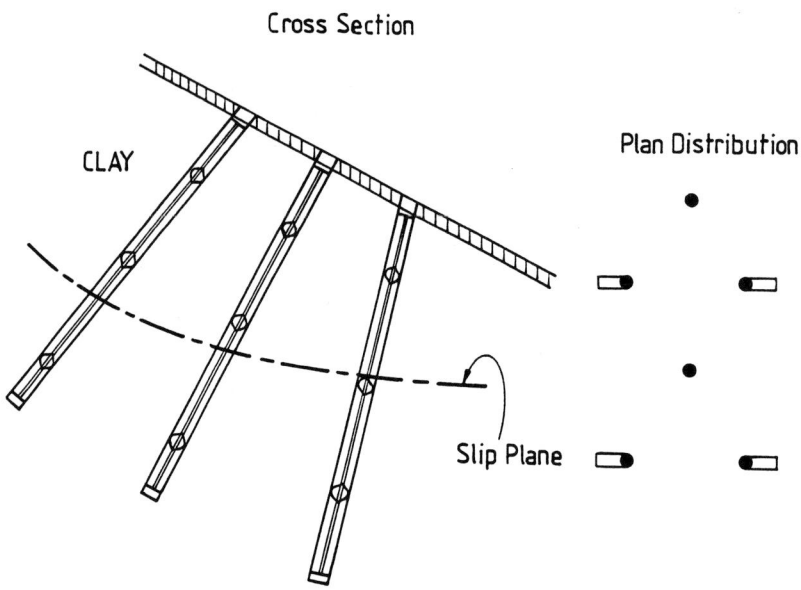

Figure 3   Typical arrangement of reinforced lime piles.

182

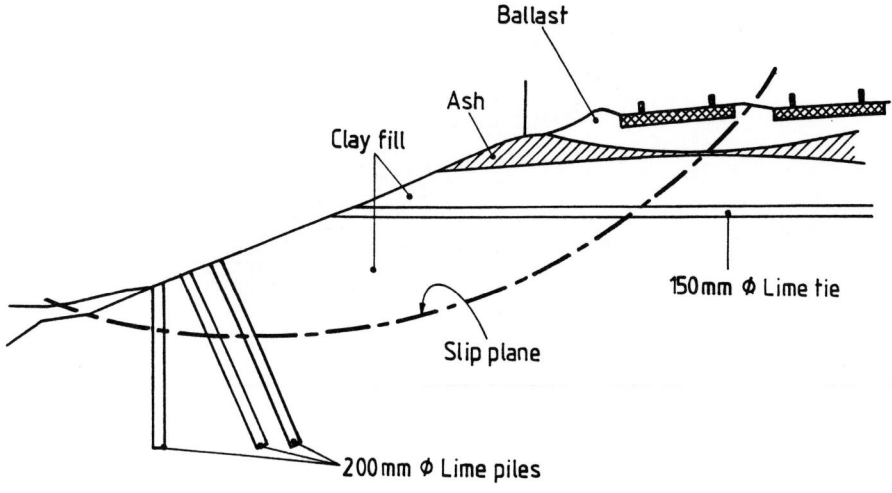

Figure 4   Typical arrangement of reinforced lime piles to stabilise embankments

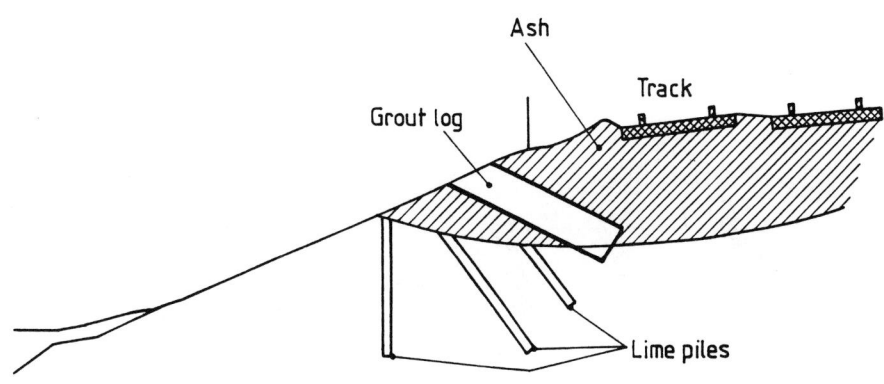

Figure 5   Typical arrangement of reinforced lime piles to stiffen embankments shoulders and enhance stability